Bats - Disease-Prone but Beneficial

Edited by Heimo Mikkola

Published in London, United Kingdom

IntechOpen

Supporting open minds since 2005

Bats - Disease-Prone but Beneficial
http://dx.doi.org/10.5772/intechopen.95729
Edited by Heimo Mikkola

Contributors
Thangavel Karuppudurai, Steffi Christiane Ramesh, Craig L. Frank, Kareem M. Soliman, Wiame W. Emam, Sarah Hooper, Sybill Amelon, Zuhair S. Amr, Mohammad Abu Baker, Omar A. Abed, Srimanta Kumar Raut, Susanta Mallick, Asif Hossain, Heimo Juhani Mikkola, Alan Sieradzki

First published in London, United Kingdom, 2022 by IntechOpen
IntechOpen is the global imprint of INTECHOPEN LIMITED, registered in England and Wales, registration number: 11086078, 5 Princes Gate Court, London, SW7 2QJ, United Kingdom
Printed in Croatia

British Library Cataloguing-in-Publication Data
A catalogue record for this book is available from the British Library

Additional hard and PDF copies can be obtained from orders@intechopen.com

Bats - Disease-Prone but Beneficial
Edited by Heimo Mikkola
p. cm.
Print ISBN 978-1-80355-012-1
Online ISBN 978-1-80355-013-8
eBook (PDF) ISBN 978-1-80355-014-5

For EU product safety concerns:
IN TECH d.o.o., Prolaz Marije Krucifikse Kozulić 3, 51000 Rijeka, Croatia, info@intechopen.com or visit our website at intechopen.com.

Meet the editor

Heimo Mikkola obtained a Ph.D. from the University of Kuopio (now Eastern Finland University), where he also served as an adjunct professor in Applied Zoology. From 1974 to 2007, he worked with the Food and Agriculture Organization (FAO) of the United Nations, first in Colombia and then in Africa, where he served as the organization's resident representative. After retiring from the FAO in Uruguay, Dr. Mikkola has worked as a part-time professor at three Kazakh universities and one Kyrgyz university. His work has taken him to 137 countries, and he has written almost 700 reports and scientific papers and books, mainly on owls and other birds, fish, insects, and food. He has studied bats for many years on almost all continents as they often share night-time activity and biotopes with owls. This is the second book on bats he has edited for IntechOpen.

Contents

Preface XI

Section 1
Bats in Folklore 1

Chapter 1 3
Bats in Folklore and Culture: A Review of Historical Perceptions around the World
by Alan Sieradzki and Heimo Mikkola

Section 2
Pesticide Exposure 23

Chapter 2 25
Pesticide Exposure Risks to Chiropteran Species and the Impacts on Emerging Zoonotic Diseases
by Sarah Hooper and Sybill Amelon

Section 3
White-Nose Syndrome 43

Chapter 3 45
The Physiological Ecology of White-Nose Syndrome (WNS) in North American Bats
by Craig L. Frank

Section 4
Ecosystem Management 61

Chapter 4 63
Bats and Ecosystem Management
by Kareem M. Soliman and Wiame W. Emam

Section 5
Dispersal Patterns and Diversity 79

Chapter 5 81
Diversity and Conservation of Bats in Jordan
by Zuhair S. Amr, Omar A. Abed and Mohammad Abu Baker

Chapter 6 **101**
Dispersal Patterns, Mating Strategy and Genetic Diversity in the Short Nosed Fruit Bat *Cynopterus sphinx* (Chiroptera: Pteropodidae) in Southern India
by Thangavel Karuppudurai and Steffi Christiane Ramesh

Section 6
Roost-Tier Preference **125**

Chapter 7 **127**
Roost-Tier Preference in Roost-Trees: A Case Study in the Bats *Pteropus giganteus*
by Susanta Mallick, Asif Hossain and Srimanta Kumar Raut

Preface

Bats are found all over the world and provide significant ecosystem services [1]. Their elusive lifestyle and unusual appearance have always stimulated peoples' imagination, perhaps more so than any other animal [2]. This book highlights how these airborne mammals have been associated with death, witchcraft, vampires, malevolent spirits, and evil in some cultures, while they have been linked to luck and good fortune and used as spiritual totems in other places.

Emerging infectious diseases (EIDs) are a growing problem for humans, wildlife, and domestic animals, despite substantial progress in disease control. It is estimated that 75% of EIDs are zoonotic, that is, they are transmitted from animals to humans. Bats are known to host the rabies virus as well as SARS, MERS, and COVID coronavirus types [3].

Due to the rabies virus in bat populations, the disease is an ever-present threat to public health that can lead to a particularly unpleasant death [4, 5]. In 1985, a bat researcher in Finland died of rabies encephalitis caused by European bat lyssavirus type 2 (EBLV-2). After that, well over 1150 bats of seven species were examined for lyssaviruses in Finland during a 28-year period. EBLV-2 may circulate in Finland even though the seroprevalence is low [6]. The health risk in Finland to the public, which has no direct contact with bats through work or leisure activities, is considered negligible [6].

Since the first case of bat rabies in Germany in 1954, 1040 cases of rabid bats have been reported in Europe [6]. This number is much higher than that in North America where millions of dollars are used "to educate" the public about the dangers of bat-borne rabies in humans [7]. This has simply exacerbated the already unreasonable fears that many people have of bats. In the period 1950–2007, only 56 cases of bat-borne rabies transmission to humans occurred in the United States and Canada, which translates to 3.9 cases per billion person-years [8], which indeed is negligible as in Finland [6].

The situation of rabies in Latin America is complex. Rabies in dogs has decreased dramatically, but bats are increasingly recognized as natural reservoirs of other rabies variants. One fifth of Latin American and Caribbean bat species have been confirmed as rabies positive. Brazil, Peru, Mexico, and Colombia are the countries with the most cases of rabies in humans in the region. In Latin America mortality rates are estimated at 0.01–0.60 per 100,000 individuals [9].

During the past decade, bats have clearly been identified as an important source of new viruses that can affect humans [10]. Bat-transmitted viruses that have caused emerging infectious diseases in humans fall into different families: paramyxoviruses including Hendra and Nipah viruses [11, 12]; Ebola and Marburg hemorrhagic fever filoviruses [13, 14]; and sudden acute respiratory syndrome-like coronaviruses [15]. This list is probably far from complete. Recently, it has been reported that avian and human influenza virus-compatible sialic acid receptors are found in little brown bats *Myotis lucifungus* widely distributed in North America, which could potentially facilitate the emergence of new zoonotic strains [16].

Epidemiologists talk about 'spillover,' which is when a virus makes the leap from one host species to another. The most dangerous spillovers to people are those from other animals to humans, creating 'zoonotic' diseases [17]. The original host of SARS-COV-2 is believed to be an as-yet-undetermined species of bat; in humans, this zoonotic disease is COVID-19 [18]. There is now broad agreement among scientists that COVID-19 probably originated in a wildlife market in Wuhan, China. The theory is that coronavirus from a bat infected wild-caught or farmed wild animals. With subsequent mutation and recombination, that virus became capable of infecting humans. A similar event was responsible for the outbreak of severe acute respiratory syndrome (SARS) in 2003 [10].

It has been shown that culling and disturbance of bat colonies have been unsuccessful in eliminating the risk of zoonotic spillover and even increased the number of infected animals in other bat-virus systems [19]. Therefore, conservationists have repeatedly emphasized the need for balanced discourse when informing the public about zoonotic risks related to bats. Even well-framed messages risk reinforcing negative associations between bats and infectious diseases, easily leading to disproportionate consequences. Thus, conservationists and health authorities are confronted with the challenge of informing people about the potential health risks associated with bats, without eroding already limited support for their conservation [19].

Additional factors that further increase the risk of disease transmission between bats and people include illegal hunting of bats for consumption and the use of bats in traditional medicine [3]. Bats have been used in medicine in various parts of the world since ancient times [20]. The oldest occurrence of bats in medicine is from 1500 BC in the medical papyri of ancient Egypt. In modern times, clinical trials are investigating the substance (desmoteplase), which is present in the saliva of the common vampire bat *Desmodus rotundus*, for its potential to help patients with acute ischemic stroke. This exploitation of animals in traditional medicine has no consideration for sustainability of populations of animals in the wild. Thus, overharvesting of medical species contributes to species loss [20].

Bat populations continue to decline worldwide because of myriad human activities. To enhance bat conservation, human behaviour needs to change. Deforestation could be a major contributing factor to new viral emergences due to more frequent contact of livestock and humans with bats possibly containing infectious viruses [10]. Monitoring bat-borne diseases and, more importantly, the environmental conditions bringing bats, viruses, and humans into contact, is crucial and should lead to the development of scenarios of risk management.

Educational campaigns should be intensified and targeted to groups that are most at risk of capturing bat-borne zoonotic diseases. Awareness programs are urgently needed in schools to improve the general knowledge of and attitudes toward bats, and to inspire the next generation of bat conservationists [1].

This book suggests that education is a suitable tool to minimize prejudice against bats and a key step to creating a harmonious coexistence between humans and bats.

Chapters address such topics as bats in folklore and culture, bat dispersal patterns, bats in ecosystem management, pesticide exposure risks, roost-tier preference, diversity and conservation, and ecology of white-nose syndrome.

My warmest thanks to Author Service Manager Nera Butigan at IntechOpen for her professional and helpful cooperation in all aspects of the publication process.

Heimo Mikkola
Eastern Finland University,
Kuopio, Finland

References

[1] Bhattacharjee J, Kumar DB, Bhattacharjee PC, Singha H, Dep P, Dutta H & Hussain MM. Student Perceptions of, and Attitudes toward, Bats in Barak Valley, Assam, India. Anthrozoös, 2018; 31(4): 411-422. DOI: 10.1080/08927936.2018.1482113

[2] Rocha R, López-Baucells A & Fernández-Llamazares A. Ethnobiology of Bats: Exploring Human-Bat Inter-Relationships in a Rapidly Changing World. Journal of Ethnobiology, 2021; 41(1): 3-17. DOI: 10.2993/0278-0771-41.1.3

[3] Mishra C, Samelius G, Khanyari M, Nuggehalli Srinivas P, Low M, Esson C, Venkatachalam S & Johansson Ö. Increasing risk for emerging infectious diseases within a rapidly changing High Asia. Ambio, 2022; 51: 494-507. DOI: 10.1007/s13280-021-01599-7

[4] WHO. Rabies Bulletin Europe. http://www.who-rabies-bulletin.org/ Accessed 12/12/2014

[5] Knobel DL, Cleaveland S, Coleman PG, Fèvre FM, Meltzer MI, Miranda MEG, *et al*. Re-evaluating the burden of rabies in Africa and Asia. Bulletin of World Health Organization, 2005; 83: 360-368

[6] Nokireki T, Huovilainen A, Lilley T, Kyheröinen E-M, Ek-Kommonen C, Sihvonen L & Jakava-Viljanen M. Bat rabies surveillance in Finland. BMC Veterinary Research, 2013; 9: 174-182

[7] Shedd W. Owls aren't Wise & Bats aren't Blind, Harmony Books, New York. 322 pp.

[8] De Serres GF, Dallaire F, Cote M *et al.* Bat rabies in the United States and Canada from 1950 through 2007 human cases with and without bat contact. Clinical Infectious Diseases, 2008; 46: 1329-1337

[9] Escobar LE, Peterson AT, Favi M, Yung V & Medina-Vogel G. Bat-borne rabies in Latin America. Journal of the São Paulo Institute of Tropical Medicine, 2015; 57(1): 63-72. DOI: 10.1590/ S0036-46652015000100009

[10] Afelt A, Devaux C, Serra-Cobo J & Frutos R. Bats, Bat-borne viruses, and Environmental changes. 2018; Chapter 8: 113-132. In: Mikkola H.(Ed.) Bats. InTechOpen, London. DOI:10.5772/intechopen.70927

[11] Halpin K, Young PL, Field HE, Mackenzie JS. Isolation of Hendra virus from pteropid bats: a natural reservoir of Hendra virus. Journal of General Virology, 2000; 81: 1927-1932

[12] Chua KB, Koh CL, Hooi PS, Wee KF, Khong JH, Chua BH, Chan YP, *et al.* Isolation of Nipah virus from Malaysian Island flying foxes. Microbes and Infection, 2002; 4:145-151.

[13] Leroy EM, Kumulungui B, Pourrut X, Rouquet P, Hassanin A, Yaba P,*et al.* Fruit bats as reservoirs of Ebola virus. Nature, 2005; 438: 575-576.

[14] Towner JS, Amman BR, Sealy TA, Reeder Carroll SA, Comer JA, *et al.* Isolation of genetically diverse Marburg viruses from Egyptian fruit bats. PLoS Pathogens, 2009; 5:e100053]

[15] Ge XY, Li JL, Yang XL, Chmura AA, Zhu G, Epstein JH, Mazet JK, *et al.* Isolation and characterization of a bat SARS-like coronavirus that uses the ACE2 receptor. Nature, 2013; 503: 535-538.

[16] Chothe SK, Bhushan G, Nissly RH, Yeh YT, Brown J, Turner G, *et al.* Avian and human influenza virus compatible sialic acid receptors in little brown bats. Nature scientific reports, 2016; 7:660/ doi:10.1038/s41598-017-00793-6.

[17] Anonymous. Covid-19 Reservoir dogs. And cats. And buffalo...The Economist November 20th, 2021: 72-73

[18] Orenstein RI. Wildlife markets, COVID-19 and totalitarianism: A comment on Cawthorn et al. Ambio, 2021; 50: 1760-1761. DOI: 10.1007/s13280-021-01568-0

[19] Rocha R, Aziz SA, Brook CE, Carvalho WD, Cooper-Bohannon R, Frick WF, Huang C-C, Kingston T, López-Baucells A, Maas B, & Mathews F. *et al.* Bat conservation and zoonotic disease risk: a research agenda to prevent misguided persecution in the aftermath of COVID-19. Animal Conservation, 2020; 24(3): 303-307. DOI: 10.1111/acv.12636

[20] Riccucci M. Bats as materia medica: an ethnomedical review and implications for conservation. Vespertilio, 2012; 16: 249-270 ISSN 1213-6123

Section 1

Bats in Folklore

Chapter 1

Bats in Folklore and Culture: A Review of Historical Perceptions around the World

Alan Sieradzki and Heimo Mikkola

Abstract

Belief systems of people have always been closely related to animals, which are symbolized in traditional narratives. Sociocultural definitions of animals as "good or evil" have persisted throughout the history of human beings. In the West, bats are often perceived as evil spirits, Vampires, and harbingers of death, while some cultures across the Asia-Pacific region associate bats with good fortune. Here, we review documented narratives and surveys from around the world and our ethnographic observations from Europe to analyze beliefs associated with bats. We explore the role that bats play in traditional narratives and the likely reasons for their salience, including their connections with the extraordinary and supernatural. Finally, we discuss shortly the need of education to change attitudes toward bats. In North America, education has had some effect as more people have started to understand how useful bats truly are and how few cases of bat-born rabies transmission to humans there have been in the United States and Canada. It remains to be seen, however, how effectively the further education efforts could halt or even reverse the decline of the bats around the world. It is also noted that bat tourism has a potential to conserve bat populations while providing social and economic benefits to local people in host communities.

Keywords: bats, folklore, culture, literature, myths, disease, need of education

1. Introduction

Bats are truly remarkable creatures, and fossil records indicate that they first appeared in the Eocene, some 50–55 million years ago [1]. They belong to the Order *Chiroptera.* This Order name means "hand-wing" as the bats can hold food between their forearms. Despite this primate-like gesture, it took quite some time for people to understand that bats are mammals and not birds. In the third book of Moses in the Old Testament, bats were identified as birds, while the world-famous Swedish taxonomist, Carl von Linné (or Linnaeus), only reclassified bats as mammals and not birds as late as 1758 in the 10th Edition of his "Systema Naturae" [2]. Because of the resemblance in dentition and such external phenomena as the thoracic position of the mammae, etc., the great Linnaeus himself ended to place the bat along with man in the order Primates [3].

Contemporarily the bats were divided based upon morphology and behavior into two suborders, Microchiroptera (Microbats) and Megachiroptera (Flying Foxes and Old World Fruit Bats) [4]. New molecular biology findings indicate that there are two new Suborders, Yinpterochiroptera and Yangochiroptera, not coinciding with the earlier subordinate classification [5].

There are more than 1400 bat species worldwide, and they make up roughly 20% of the world's extant mammals [6]. They are hugely beneficial to man and play a major role in the well-being of the world's ecosystems. Not only do they prey upon insects that are harmful to agriculture, but they also prey upon mosquitoes and other virus-carrying insects and play a major role in pollinating and spreading the seeds of many of the fruits we enjoy. In some parts of the world, they are a valuable food source, and their body parts are used in traditional medicine, while their guano is collected and used as agricultural fertilizer [7]. Why is it then that while in some parts of the world this beneficial creature is seen as a symbol of good luck and good fortune, in many societies it is viewed with fear and loathing? Here, using documented narratives, surveys, popular literature, and cinema, we will explore the various myths, legends, and attitudes to bats from around the world.

2. Europe

In the Bible, the bat is seen to be "unclean" [8], while its nocturnal activities ally it to malevolent spirits that roam the land when darkness has fallen. It is no real surprise that in a Christian Europe throughout history, the bat has been associated with the Devil, evil spirits, and witches [9]. Bats also have wings. Tertullian, an early Christian author from Carthage (155–220), claimed that the Devil and his angels had wings [10], and around 1314, Dante wrote that the Devil's wings had no feathers, "but was in form and texture like a bat's" [11]. In 1332, a French noblewoman, Lady Jacaume of Bayonne [12], "was publicly burned to death as a witch because 'crowds of bats' were seen about her house and garden."

William Shakespeare (1564–1616) also equated bats with witches, spells, and curses. In *Macbeth* (1605), there is the incantation of the three witches: "Eye of newt, and toe of frog, wool of bat, and tongue of dog," while there is Caliban's curse on Prospero in *The Tempest* (1610–1611): "All the charms of Sycorax, toads, beetles and bats, light on you."

There is also, of course, the European connection with bats to vampirism. Vampires had been part of Slavic folklore in Eastern Europe since the seventeenth and eighteenth centuries, but it was not until the nineteenth century that popular fictional literature, predominantly through Bram Stoker's "Dracula" in 1897, would forever tie bats and vampires together, with the protagonist, Count Dracula, being able to transform himself into a huge bat [13]. While the three species of true vampire bats (common vampire bat *Desmodus rotundus*, hairy-legged vampire bat *Diphylla ecaudata*, and white-winged vampire bat *Diaemus youngi*) are microbats, measuring just a few centimeters in length, it was the exaggerated reports from early explorers and adventurers that gave the public the image of these huge blood-sucking creatures. In 1796, John Stedman wrote of being bitten by a vampire in Guiana, describing it as "a bat of monstrous size, that sucks the blood from men and cattle when they are fast asleep, even sometimes till they die." [14]. This association between giant bats and vampires continues to the present day through popular fiction and cinema (**Figure 1**).

As with that other iconic nocturnal creature, the owl, the bat has a myriad of very strange old wives' tales and superstitions surrounding it [15]. Here are just a few:

If a bat flies into your house, look out for bedbugs.

A bat flying into a building means that is going to rain.

It's unlucky to see a bat in the daytime.

Figure 1.
An old engraving of vampire bats described by early explorers such as John Stedman. Public domain. Courtesy of "creative commons"—Wikimedia.

Killing a bat shortens your life.

Bats in a church during a wedding ceremony is a bad omen.

Bats in the house mean a death in the house or is a sign that the occupants will soon be leaving.

Bats flying vertically upwards and then dropping back to earth means that the Witches Hour has come.

Bats are symbolic of bad luck, especially when they call while flying early in the evening.

If a bat flies into a house and then escapes, there will be a death in the family. Kill the bat before it escapes, and everyone will be safe.

If a bat flies into a kitchen and at once hangs on to the ceiling, it is a lucky omen, but if it circles the room twice before alighting, it is a bad omen.

One of the most enduring old wives' tales from Europe is that bats will get tangled in women's hair and would have to be removed with a pair of scissors. Between 1958 and 1961, Gathorne-Hardy, Fifth Earl of Cranbrook, a renowned conservationist and a founding member and former President of the Mammal Society, decided to put this old superstition to the test [16]. Using two willing female teenagers, *one with "relatively short curly hair" and one with "longer wavy hair was done up behind in a bun,"* Cranbrook took turns in placing four different species of bat on their heads. A noctule bat *Nyctalus noctula*, a long-eared bat *Plecotus auritus*, a Natters bat *Myotis nattereri* and a Daubenton's bat *Myotis daubentonii*. In all four cases, the bats each walked about the volunteers' hair without becoming entangled in any way and finally took flight without any difficulty. The experiment was repeated several times with the same results [16].

3. Africa

Bats have fascinated humans for millennia, and this cultural and spiritual relevance is reflected in the presence of bat symbols in Egyptian tombs from 2000 BC [7]. In Africa, "house bats" occur in distinct colors and sizes, but they are usually hard to identify. Although they live near humans and are common, it appears that some of the species have not been described and given scientific names [17]. To date, ethnobiological information regarding bats in Africa has mainly focused on utilitarian aspects of bats as food and medicine, whereas knowledge concerning symbolism and beliefs surrounding bats in that continent has not been assembled, apart from ancient Africa [18], Ghana [19, 20], Kenya [21], and Madagascar [22].

Among the Ibibio people of southern Nigeria, bats are associated with witchcraft, and for any bat to fly into a house and touch a person is a sure sign that this person is thereafter bewitched and will soon perish because his or her heart is eaten at night while he or she sleeps. Also, in Nilotic Sudan, witchcraft was usually performed at night, and therefore, owls and bats were associated with it. In the Anglo-Egyptian Sudan, spirits, devils, and witches had their general name "Bitaboh," wood-goblins being specially called "Ronga." All the bats were comprehended under the same name, especially the Yellow-winged Bat *Lavia frons*, formerly *Megaderma frons*, which flutters about from tree to tree in broad daylight [18].

In Cameroon, the superstition of the vampire is attached to bats. Among the Ndop people, bats, owls, and bush-cats are said to be witch shapes. Should a bat or an owl come near the house, or a bush-cat defecate in the compound, the owner must go at once to diviner to discover what remedies must be taken to ward off the evil. A witch shape is believed to be capable to suck out the life of a sleeping man or woman [18].

From Sierra Leone comes an account of the gruesome habits of the Hammer-headed Fruit Bat *Hypsignathus monstrosus*. It is the largest bat found in continental Africa and was believed to suck the blood of sleeping children until they die. It was called "Boman," and it was able to turn into a stone or a snake at will. Interestingly, blood sucking has been attributed to bats both in Cameroon and Sierra Leone despite the countries being widely separated and when no such type of bat is found in Africa [19].

Although the bats roost in the sacred forest in Ghana, they are not regarded as sacred animals. Bat hunting is illegal, but hunters readily admitted to having hunted bats and even directed the research people to other hunters. Bat meat is widely consumed because it is considered more delicious than other types of meat. However, consumption was influenced by religious beliefs, food taboos, and some myths about bats. Muslims and Seventh Day Adventists did not consume bat meat. Men hunt and consume bats more often than women who have fears that consuming bat meat would give them strange or deformed children. Bats are not associated with any diseases, and it was felt to be safe to eat bat meat, but both studies were conducted before the 2013 Ebola outbreak (**Figure 2**) [19, 20].

In Kenya, a bat attitude questionnaire was presented to 394 people living around the Arabuko-Sokoke forest. Belief in myths seemed to prevail among those surveyed. Just over one-third of the respondents (36%) did not see any benefits of bats to humans. Nearly another third reported actively killing bats or destroying bat roosts, and most respondents associated bats with the destruction of farmers' fruits, especially mangoes. Female respondents in this study showed more negative attitudes and a stronger belief in myths about bats than males. Only older and somewhat educated people reported more positive attitudes toward bats than others [21].

In Madagascar, the exploitation of bats for bush meat regularly takes place during periods of food shortage, especially fruit bats, which are heavily hunted. The study estimated that in the karstic Mahafaly Plateau some 50–100 caves are exploited for bats and that between 70,000 and 140,000 Microchiroptera bats

Figure 2.
Large-bodied old-world fruit bats, like this Egyptian fruit bat Rousettus aegyptiacus, *are disproportionally targeted to be hunted for bushmeat. Photo credit to "creative commons" https://creativecommons.org/licenses/by-sa/4.0/legalcode.*

may be collected annually in the region. This is bound to threaten the continued existence of local bat populations. Thus, more famine relief food aid is desperately needed to reduce the bush meat use during the food crises, which are nearly annual in the region [22].

In Malawi, negative bat superstitions caused recently quite disproportionate behavior when rural people in southern parts of the country killed nine people accusing them to be "vampire bats" [23]. The police arrested 200 vigilante youth suspected of involvement in gangs that attacked persons allegedly engaged in vampirism. Medics said there is no truth or clinical evidence that blood suckers or vampires exist in Malawi, and the United Nations mission withdrew its personnel from the riots-affected areas [24].

4. Southwest Asia

In southwest Asia, bat folklore has been documented between Iran, India, and Myanmar [25–27]. In southwest Asia, the bat is predominantly perceived as a strange, negative, and demonic animal that should be avoided. Usually, it is regarded as a bad omen even if seen in a dream. The bat is a devil who flies only in the night because it would die if flying in the daytime [25].

In India, a total of 3059 high school students from 36 schools were questioned on their bat perceptions. Most students (56%) reported seeing bats in their locality. Half of the students knew that bats are mammals, but 26% believed them to be birds, some thought of them being amphibians (12%) or reptiles (11%); 37% disliked the bats, and 27% liked them remaining seeing them as indifferent. A substantial proportion (53%) thought that bats have medicinal value, but 35% saw them only as harmful creatures. They were believed to destroy trees and damage fruits in gardens. Over half of the students (57%) did not know anybody in their neighborhood who would hunt bats. The felling of roosting trees was mentioned as bat death reason by 29% and hunting by 18%. Most students wanted to participate in bat conservation activities, including some students who disliked bats. The study concluded that conservation-related education should be included in the school curriculum to ensure that students would appreciate not only the importance of bats but of all wildlife [26].

Bats are also revered throughout India. In Madurai, worshippers of the Muni god regard Indian Flying Fox *Pteropus medius,* formerly *Pteropus giganteus,* as sacred and protect colonies for fear of heavy punishment. However, after offering prayers, dead bats found on the ground can be eaten. In Pudukkottai, roosting *P. medius* is seen as the guardian of the sacred groves, and in Bihar, that bat brings wealth. Orchard owners in Myanmar believe that allowing bats to roost and feed in the garden ensures prosperity and well-being. Flying fox's emergence time is believed in Myanmar to indicate the weather: an early or no emergence foretells a coming storm. Bats are also used as allegories to denote romantic or parental love in the poetry of Tamil Sangam literature in India [27–29].

In Hunza, Pakistan, a witch appears in the shape of a bat, and it plays a role in magical practice in southwest Asia. In the Punjab, northern India, magicians use bat bones to prepare their concoctions, and with the bat blood, they write amulets for malevolent and antisocial magic. Between Iran and Rajasthan, north-western India, there is the widespread public belief that if a bat lives somewhere in the house or hangs on the roof, this would mean bad luck for the owner [25]. In Pakistan, one belief is that if a bat enters one's ear, it can never be removed and contact with bat urine is thought to cause eczema. On the other hand, the body fat of Indian Flying Fox is used to make massage oil to cure rheumatic pains, while drinking water from a bat's wing is said to sharpen one's memory [30]. In Sri Lanka, it is believed that one may be reincarnated as a bat for denying another person drinking water [31].

Pre-Islamic magical belief in Hunza said that "if the lady would once offer the cooked meat of bat to her guests, hiding it behind her back while serving" her family would never be without meat for their whole life [25]. Interestingly, Islamic taboo,

considers bat as *harām* meaning that it is forbidden as food. The official Islamic view is exempting bats from being killed because it would carry bad luck to kill a bat [32]. One reason not to kill bats is that "female bats have breasts and are mothers like human females" [25].

Bats are also seen as useful in Iran, India, and Pakistan because people are collecting bat guano as a natural fertilizer.

5. Southeast Asia

Throughout Southeast Asia, bats are associated with luck and good fortune and used as spiritual totems [6, 27]. Since the fourteenth century, Chinese culture has regarded bats as lucky animals, and these blessing bat symbols have been prevalent in

Chinese arts over the centuries [33]. In Indonesia, farmers in South Sulawesi believe that flying foxes roosting near their rice fields guarantee a good harvest [27]. Similarly, fishermen in the Philippines consider mangrove roosting flying foxes to be guardians of their fishing grounds and to increase fish and shellfish catch [27]. In Malaysian Borneo, people consider it taboo to disturb a fruit bat. If a man whose wife is nearing childbirth unthinkingly does so, some harm may befall the unborn baby [34].

In Malaysia, a face-to-face survey was conducted in Penang Island to assess knowledge and awareness level toward bat conservation efforts. The bat populations in Malaysia are decreasing in 26% of species and only 15% are still stable. Bats are shot for sport or to eradicate them from fruit plantations. Bats are also considered as wild exotic meat, which is widely consumed in urban areas. Besides hunting, the primary threats to bat species include habitat loss and degradation through logging. Most respondents were less likely to value the importance of bats in the ecosystem, so creating a conservation education to connect people with nature is not easy. The oldest age group (51–70 years old) of 150 respondents were mostly aware of the bat conservation efforts. The higher level of education was not always reflected in the positive attitudes toward environment and wildlife issues. Participation by local people is vital to achieving successful conservation programs [35].

Iban people in Sarawak, Borneo, believe that a bat flying into the house indicates a shaman bringing good vibes, conferring protection against any harm, while in Thailand, if a bat enters a house but immediately flies away, it is believed to change bad luck to good. Should the bat stay and eat the fruit in the house, bad luck will befall the owner [27].

In northern Thailand, harming bats incurs a curse because bats are sacred for Buddhists [36]. In Irian Jaya of Indonesia, former head-hunters considered flying foxes to be head-hunters too, as they took the "head" of the tree by consuming its fruit [37].

Despite positive associations toward bats, they are widely consumed as food and medicine throughout Asia, except in Brunei and Singapore. In Malaysia, ethnic Han Chinese, non-Muslim indigenous groups, and ethnic Malays hunt flying foxes and trade them to the Chinese [27]. In Indonesia, Iban people in Kalimantan also hunt

Figure 3.
Bats for eating in the Laos marketplace. Photo credit: Stan Delone "creative common"—Wikimedia. https://creativecommons.org/licenses/by-sa/2.0/legalcode.

fruit bats for consumption, and mainly Christian people in North Sulawesi regularly eat flying foxes (Black Flying Fox *Pteropus alecto* and Sulawesi Fruit Bat *Acerodon celebensis*) especially during the Christmas season [38, 39].

There exists a widespread belief in Southeast Asia that eating bat meat cures asthma [27]. In many parts of Indonesia, people specifically consume bats' livers and hearts as medicine [39]. In Thailand, bat meat or blood is eaten for muscle pain, increasing virility and longevity [36]. In Malaysia, older generations prevented thievery by mixing flying fox blood with milky mangrove *Excoecaria agallocha* tree sap to cause violent intestinal inflammation [40].

In Vietnam, many mounted bat species are sold in souvenir shops, and in Laos, bats are traded in several markets (**Figure 3**) [41, 42]. There is a Japanese word for bat, *komori*, which is said to mean "mosquito slaughterer." In Japanese mythology, very old bats can transform into *nobusuma*, spirit animals resembling flying squirrels that land on their victims' faces at night to feed off blood [43]. Indigenous Ainu people in Japan worshipped the crafty and wise bat god *Kappa kamui,* who kept away demons and diseases [44].

6. Pacific region

A legend tells that in Fiji, a giant white vampire bat acts as a messenger, and one finds bats as gods in both Tikopian and Tongan myths. The Tongan king's Samoan wife was rescued by flying foxes, and she honored her rescuers later by naming her son Tonumaipe'a (= "rescued by flying foxes") [45]. People in Vanuatu consider Pacific Flying Fox *Pteropus tonganus* as their ancestor and claim to be able to communicate with them [46]. In Makira, Solomon Islands, local people value traditional currency for transactions, such as bride price, and use the canine teeth of flying foxes as a traditional currency [47]. Samoan people prize flying fox meat as a delicacy and as a gift to elders, but commercial hunting and export of the meat are culturally frowned. The general attitude is that the flying fox is part of the forest, and the vast majority support the protection of *Pteropus samoensis* and *P. tonganus* [45]. Samoans said that flying foxes were cheeky and courageous, making it a popular tattoo motif [48].

In New Zealand, Māori people associate bats, *pekapeka*, with the mythical nocturnal bird *hokioi* that foretells death [49]. In 2021, a bat won New Zealand's Bird of the Year competition name of which in Māori language is *Te Manu Rongonui o te Tau*, and the word "*Manu*" means "flying creatures," including bats. The decision to include the New Zealand long-tailed bat *Chalinolobus tuberculatus* also known as *pekapeka-tou-roa* (Māori) in the 2021 Bird of the Year competition did cause a bit of controversy, some people saying the country had gone "batty." However, the long-tailed bat got more than 7000 votes, bringing a clear victory to this critically endangered animal, despite not being a bird [50].

7. Central and South America

The diversity of South American bats is impressive as there are more bats, and more bat species, than in any other part of the world [51, 52]. In pre-Columbian Central and South America, the bat played an important role in the religions and social structures of the various cultures, most notably with the Moche people of Peru and the Maya of Guatemala. In northern Argentina, a Toba story tells of the leader of the very first people—a hero bat or batman who was teaching people all they needed to know as human beings. Similarly, the Ge tribe in Brazil moved through the night following a bat that looked for light toward which to guide the people (**Figure 4**) [51, 52].

Figure 4.
A bat-headed figure from Costa Rica made by pre-Colombian Diquis people sometimes between 700 and 1530 AD. Photo credit: Public domain "creative commons"—Wikimedia.

The bat was central to Maya religion and social structure. One clan of the Cakchiquel Maya, of the highlands of Guatemala, was named the Zotzil (=belonging to the bat), whose deity was a bat. The Tzotzil Maya lived, and continue to live to this day, on the plateau of Chiapas in southern Mexico. They called themselves Zotzil uinic (batmen), claiming that their ancestors discovered a stone bat, which they took as their god, and their chief town was named Zinacantlan (=place of the bats) by Nahuatl merchants from Mexico [53]. Generally, the Maya revered a Vampire Bat god, Camazotz, the death bat, which killed dying men on their way to the center of the earth [51, 52].

The north coast of Peru is one of the South American regions where bat iconography is particularly prominent [51]. The Moche people in Peru were aware of the connection between bats and plants. On Mochica pottery, a bat is depicted with the Sweetsop *Annona squamosa*, a common fruit also known as Sugar Apple or Pinha, the seeds of which are dispersed by bats [51, 52]. Some of their ceramic vessels have an anthropomorphic bat that is an agent of human sacrifice, with a knife in one hand and a human head in the other. Sometimes a Mochica bat carries a warclub and a small human captive. The enormous size of the bat and the small human head or body indicate supernatural status for the bat [51].

The widespread sacrificial association derives largely from the habits of the Common Vampire Bat that feeds exclusively on the blood of vertebrates (**Figure 5**).

In many places, blood sacrifice was believed to benefit agriculture, and therefore, bats had agricultural, as well as death, connections for Pre-Columbian peoples and in Oaxaca, Mexico, a bat deity was associated with maize [51]. Surprisingly little folklore exists specifically about Vampire bats, and Pre-Columbian erotic scenes do not involve bats although some folklore portrays female bats as alluring to men.

One Yupa man in northern Colombia started night after night to drink and flirt with a female bat when he was returning from an evening hunt. Finally, his wife realized what he was doing and set fire to the tree and killed her husband and the bats [54]. Sometimes bats are husbands as in a Mataco lore from Argentina. A woman noted that her husband had a round tail and dropped the vessel of water she was bringing to him. The bat husband then cut off her head and those of other

Figure 5.
Common vampire bat Desmodus rotundus *is the one reason for global Chiroptophobia although it is very small weighing about 55 grams and only occurring in central and South America. Photo credit: Uwe Schmidt. "Creative commons"—Wikimedia. https://creativecommons.org/licenses/by-sa/4.0/legalcode.*

Indians and put all heads in the tree hole where he was living [55]. Also, a Tacana woman in Bolivia was killing a bat while not realizing that it was her husband [34].

In some folklore, bats often have sexual connotations, which may relate to fertility and agriculture as bats are important seed dispersers and pollinators of many fruit trees. Bat guano provides one reason for the fertility associations [56].

In Caribbean South America and the Antilles, bat images are associated with death rites and burials in archeological context [57], and in Cuba, a Taino ball court was bat-shaped, the ballgame being a sacrificial ritual [58]. In Jamaica, the bat and the owl were very important symbols in Taino mythology and death. The bat represented the *opias* (= spirits of the dead people) to the Taino. Fruit-eating bats such as Jamaican fruit bat *Artibeus jamaicensis* loves feeding on guavas, which is also the favorite food of the Taino spirits of the dead. In Jamaican folklore, bats are also perceived as death images [59].

In northern Guiana, Bat Mountain is the home of" killer bats," and there is also a killer bat in folklore from Venezuela. Decapitating bat demons appears in various myths in Amazonia and to the south in northern Argentina. These myths associate killer bats with fire as the bat burns its victims and is, in turn, thrown into the fire [60–62]. One bat, whose habits may have fostered these tales of decapitating bats, is the false vampire *Vampyrum spectrum.* It is the largest New World bat with a yard wingspan. It is a carnivore, eating birds and other vertebrates, occasionally taking even other species of bats. When capturing its prey, it grabs the neck, sometimes killing the prey with a single powerful bite [51, 52].

8. North America

The artificial bat became a shorthand for horror in 1931, jiggled on a fishing line behind a Hungarian-American actor Béla Ferenc Dezsö Blaskó, known professionally as Bela Lugosi in the genre-defining movie *Dracula.* Shockingly, US news media recently reported shortages of Halloween decorations—plastic bats among them, doubtless—due to the world supply crunch [63].

Monstrous beyond imagining, all-consuming, blacker than blackest night, the hideous Satan in the Night on Bald Mountain section of Walt Disney's animated film *Fantasia* (1940) spreads gigantic bat wings as it turns fiery eyes toward the lost souls about to be engulfed in wrath and flames [64]. This North American bat-like depiction should come as no surprise; it is merely another manifestation of the fear, horror, and superstition with which bats have been regarded down through ages [64]. Large, often man-eating, bats are found in Hawaiian traditions, and there is even a legend of an eight-eyed bat in Hawaii. Interestingly, giant bat stories have persisted in these" flying-fox-less" societies, which evolved from western Polynesians cultures where flying foxes have been prevalent [45].

Warner Shedd (2000) felt strongly that the level of fear about rabid bats sometimes rises almost paranoid concern in the United States, citing the State of New York as an example [63]. By using a million dollars annually "to educate" the public about the dangers of bat-caused rabies in humans simply exacerbates the already unreasonable fears, which many people have of bats. In its entire history, the state of New York has recorded only one case of bat-transmitted rabies [64] Between the years 1950 and 2007, only 56 cases of bat-borne rabies transmission to humans occurred in the United States and Canada, which translates to 3.9 cases per billion person-years [65].

The recent introduction of a fungal disease (WNS = White-nose Syndrome) from Eurasia to North America has killed millions of bats in North America in the past decade. Although the exact source of the fungal pathogen, *Pseudogymnoascus destructans*, and its mode of introduction into North America remain unknown, the introduction was most likely mediated by humans, either through direct or indirect transfer of infectious propagules [66]. People can further move the fungus on their clothing and caving gear and spread the disease into an area that does not currently. have the fungus [67].

Even worse have been the vandalism and wanton destruction of bats and their habitat in North America. A variety of methods have been used to harass and kill these harmless and beneficial creatures, and some people have even gone so far as to dynamite caves and abandoned mines where bats roost or hibernate [64]. Lately, education seems to have some effect, and more and more people have started to appreciate how useful and amazing bats truly are. It remains to be seen, however, if this could halt or even reverse the decline of the North American bats [64].

9. Fear of bats

An irrational fear of bats—Chiroptophobia—encompasses negative perceptions of bats as disease vectors, pests, or harmful creatures associated with devils and witchcraft, which represents an important barrier to bat conservation globally [27]. Especially Western cultural associations of bats with evil spirits, which have been recorded in Christian tradition as early as the fourteenth century [68] along with current media sensationalizing bats and the COVID-19 pandemic, are major hindrances toward bat conservation [69], although there is no reason why public health messages cannot be consistent with bat conservation.

In Western culture, bats are also associated with vampires causing hysteria and wrong nomenclature of non-sanguivorous bat species such as Large Flying Fox *Pteropus vampyrus*. Despite its scientific name, it feeds exclusively on fruits, nectar, and flowers and not blood (**Figure 6**) [70].

Since the fourteenth century, Chinese culture has associated bats with good luck and blessings [33].

Even cultures that value bats positively may have values conflicting with bat conservation. Fruit growers may view fruit bats as crop pests [71]. Han Chinese people attach positive esthetic values to bats but may still hunt bats at unsustainable levels [72]. In India, bats are revered in many areas, but still large bat-harvesting festivals take place. In Nagaland between 7000 and 25,000 cave-dwelling bats (Cave Nectar

Figure 6.
Large flying fox Pteropus vampyrus *suffers from its misleading scientific name as it is not blood-eating species but feeds only fruits and flowers. Photo credit: Masteraah. "Creative commons"—Wikimedia. https://creativecommons.org/licenses/by-sa/2.0/de/legalcode.*

Bat *Eonycteris spelaea*, Great Roundleaf Bat *Hipposideros armiger*, and Leschenault's Rousette *Rousettus leschenaultii*) have been annually harvested until recently [73]. So, using positive symbolism alone to promote bat conservation might be insufficient without accounting for day-to-day relationships of local communities with bats. Conversely, negative symbolism could promote conservation if it prevents unsustainable hunting and consumption [27].

It is imperative to obtain more current ethnobiological data to further our understanding of contemporary attitudes and relationships with bats—and to document other cultural traditions not covered in this review.

10. Conclusion

Cultural traditions and beliefs influence the future of the bats, which is threatened by human exploitation, both directly on bats and indirectly on the environment.

To summarize shortly the various findings: In the West, bats have been seen as animals of ill omen, alongside other nocturnal hunters such as owls and black cats. Around 62% of the Asia-Pacific people had only positive cultural values of bats, 8% had only neutral values, while 10% had only negative values [27]. The remaining cultures had combinations of positive, neutral, and negative values.

This suggests that the Asia-Pacific region and its cultures contain far more positive associations with bats than most European or American societies and, as such, offer promising examples and opportunities to promote human-bat coexistence. For example, a number of these countries are investing in the burgeoning industry of ecotourism by promoting organized visits to bat caves and "bat watching" [74, 75]. In 2005, **the privately owned** Montfort Bat Cave Sanctuary on Mindanao Island, south of the Philippines, opened for tourism. The five-chambered cave hosts the world's largest known colony of the Geoffroy's Rousette Fruit Bat, *Rousettus amplexicaudatus*, with an estimated 1.8–2 million individuals [76]. The tour includes a brief

20-minute conservation education lecture about bats, caves, and the history of the Montford Bat Cave Sanctuary. The cave prohibits hunting and guano harvesting, while income from this ecotourism chiefly flows into local economy and enhances the sustainability and protection of the cave site (**Figures** 7 and **8**).

Our collection of global perceptions aims to promote a better biocultural richness for humans and bats as our long-term nocturnal companions. But it became obvious that the public attitude toward bats has still not been investigated extensively enough throughout the world, namely in contemporary Africa, America, Australia, and Europe.

Figure 7.
Montfort bat cave entrance in the Philippines with the numerous Geoffroy's Rousette fruit bats Rousettes amplexicaudatus. *Photo credit: [7] Roy Kabanlit "creative commons"—Wikimedia https://creativecommons.org/licenses/by-sa/4.0/legalcode.*

Figure 8.
A close-up photo of the Geoffroy's Rousette fruit bats Rousettes amplexicaudatus *at the Montfort bat cave entrance. Photo credit: Raniel Jose Castaneda "creative commons" – Wikimedia credit as in* ***Figure*** *7.*

Author details

Alan Sieradzki[1] and Heimo Mikkola[2*]

1 The Global Owl Project Carriage Drive, Alexandria, VA, USA

2 Eastern Finland University, Kuopio, Finland

*Address all correspondence to: heimomikkola@yahoo.co.uk

References

[1] Simmons NB, Seymour KL, Habersetzer J, Gunnell GF. Primitive early Eocene bat from Wyoming and the evolution of flight and echolocation. Nature. 2008;**451**:818-821. DOI: 10.1038/nature06549

[2] Linnaeus C. Systema Naturae. 10th ed. Vol. 1. Stocholm: Laurentius Salvius; 1758. pp. 1-824

[3] Oliphant SG. The story of Strix ancient. Transactions and Proceedings of the American Philological Association. 1913;**44**:133-149

[4] Richardson P. Bats. Stansted: Whittet Books; 2017. 138 pp

[5] Lei M, Dong D. Phylogenomic analyses of bat subordinal relationships based on transcriptome data. Scientific Reports. 2016;**6**(1):1-8. DOI: 10.1038/srep27726

[6] Rocha R, López-Baucells A, Llamazares F. Ethnobiology of bats: Exploring human-bat InterRelationships in a rapidly changing world. Journal of Ethnobiology. 2021;**41**(1):3-17

[7] Kunz TH, Braun de Torrez E, Bauer D, Lobova T, Fleming TH. Ecosystem services provided by bats. Annales New York Academy of Science. 2011;**1223**:1-38. DOI: 10.1111/j.1749-6632.2011.06004.x

[8] Leviticus 11:19 in Old Testament, The Bible. Available from: https://www.kingjamesbibleonline.org/Leviticus-11-19/

[9] Eklöf J, Rydell J. Attitudes towards bats in Swedish history. Journal of Ethnobiology. 2021;**41**(1):35-52. DOI: 10.2993/0278-0771-41.1.35

[10] Tertullian. Apology and De Spectaculus. Minucius Felix: Octavius. Translated by Glover TR & Rendall GH. Cambridge, Massachusets: Harvard University Press; 1931. 121 pp

[11] Dante A. Dante's Inferno. Translated by Musa M. Bloomington, Indiana: Indiana University; 1995

[12] Cooke L. Unexpected Truth about Animals. New York: Random House; 2017. 480 pp

[13] Dodd K. "Blood Suckers Most Cruel:" the vampire and the bat in and before Dracula. Athens Journal of Humanities & Arts. 2019;**6**(2):107-132

[14] Stedman J. Narrative, of a Five Years' Expedition, against the Revolted Negroes of Surinam, in Guiana, on the Wild Coast of South America; from the Year 1772, to 1777: Elucidating the History of that Country, and Describing its Productions, Viz. Quadrupedes, Birds, Fishes, Reptiles, Trees, Shrubs, Fruits, & Roots; with an Account of the Indians of Guiana, & Negroes of Guinea. Vol. 2. London: Johnson J and Edwards J; 1796. pp. 142-144

[15] Daniels CL, Stevens CM, editors. Encyclopedia of Superstitions, Folklore and the Occult Sciences of the World. Vol. II. University Press of the Pacific; Oregon, Forest Grove: Pacific University; Oregon, Pacific University: Forest Grove; 2003. 668 pp

[16] Cranbrook E. Bats in Women's hair. Transactions of the Suffolk Naturalists' Society. 1958-1961;**2**(1-6):193

[17] Eklöf J, Rydell J. Bats In a World of Echoes. Cham, Switzerland Springer; 2018

[18] Jeffreys MDW. African pterodactyls. Journal of the Royal African Society. 1944;**43**(171):72-74

[19] Ohemeng F, Lawson ET, Ayivor J, Leach M, Waldman L, Ntiamoa-Baidu Y.

Socio-cultural determinants of human-bat interactions in rural Ghana. Anthrozoös. 2017;**30**(2):181-194. DOI: 10.1080/08927936.2017.1310984

[20] Anti P, Owusu M, Agbenyega O, Annan A, Kofi Badu E, Nkrumah EE, et al. Human-bat interactions in rural West Africa. Emerging Infectious Diseases. 2015;**21**(8):1418-1421

[21] Musila S, Prokop P, Gichuki N. Knowledge and perceptions of, and attitudes to, bats by people living around Abuko-Sokoke Forest, Malindi Kenya. Anthrozoös. 2018;**31**(2):247-262

[22] Goodman SM. Hunting of Microchiroptera in South-Western Madagascar. Oryx. 2006;**40**(2):225-228. DOI: 10.1017/S0030605306000354

[23] Bambi J. Malawi 'Vampire' Rumours and Hunt. The Morning Call. Available from: https://www.africanews.com/2017/10/30/malawi [Accessed: 10/12/2021]

[24] Anonymous. In Malawi, the "Vampire Hunters". The Earth Chronicles of Life, Oct 2017. Available from: https://earth-cronicles.com/anomalies/in-malawi-the-vampire-hunters.html [Accessed: 10/12/2021]

[25] Frembgen JW. Embodying evil and bad luck. Stray notes on the folklore of bats in Southwest Asia. Asian Folklore Studies. 2006;**65**:241-247

[26] Bhattacharjee J, Dutta BK, Bhattacharjee PC, Singha H, Deb P, Dutta H, et al. Student perceptions of, and attitudes toward, bats in Barak Valley, Assam, India. Anthrozoös. 2018;**31**(4):411-422. DOI: 10.1080/08927936.2018.1482113

[27] Low M-R, Hoong WZ, Shen Z, Murugavel B, Mariner N, Paguntalan LM, et al. Bane or blessing? Reviewing cultural values of bats across the Asia-Pacific region. Journal of Ethnobiology. 2021;**41**(1):18-34

[28] Marimuthu G. The sacred flying fox of India. BATS Magazine. 1988;**6**:2

[29] Tangavelou AC, Rani PJ, Karthikeyan S. Conservation of sacred Indian flying fox (bat) at sacred landscape of Pudukottai District, Tamil Nadu, India. Asian Journal of Conservation Biology. 2013;**2**:178-180

[30] Mahmood-ul-Hassan M. Faiz-ur-Rehman, Salim M. public perceptions about the fruit bats in two horticulturally important districts of Pakistan. Journal of Animal and Plant Sciences. 2011;**21**(1):135-141

[31] Simon GH. Ceylonese beliefs about animals. Western Folklore. 1954;**13**:260-267

[32] Eisenstein H. Einfűhrung in die arabische Zoographie. Das tierkundliche Wissen in der arabisch-islamischen Literatur. Berlin: Dietrich Reimer; 1991

[33] Eberhard W. A Dictionary of Chinese Symbols: Hidden Symbols in Chinese Life and Thought. London: Routledge; 1986

[34] Metcalf P. Supernatural Etiologies of illness in central northern Borneo. The Malaysian Branch of the Royal Asiatic Society. 1982;**55**:115-125

[35] Shafie NJ, Mohd Sah SA, Mutalib AHA, Fadzly N. General perceptions and awareness level among local residents in Penang Island towards bats conservation efforts. Tropical Life Sciences Research. 2017;**28**(2):31-44

[36] Suwannarong K, Balthip K, Kanthawee P, Khiewkhern S, Lantican C, Ponlap T, et al. Bats and belief: A sequential qualitative study in Thailand. Heliyon. 2020;**6**:e04208. DOI: 10.1016/j.heliyon.2020.e04208

[37] Smidt D. Asmat Art: Woodcarvings of Southwest New Guinea. George Braziller in Association with Leiden. New York: Rijksmuseum voor Volkenkunde; 1993

[38] Wadley RL, Colfer CJP. Sacred Forest, hunting and conservation in West Kalimantan. Human Ecology. 2004;**32**:313-338

[39] Sheherazade TSM. Quantifying the bat Bushmeat trade in North Sulawesi, Indonesia, with suggestions for conservation action. Global Ecology and Conservation. 2015;**3**:324-330

[40] Gimlette JD. Malay Poisons and Charm Cures. 3rd ed. London: J. and A. Churchill; 1929

[41] Francis CM, Guillén A, Robinson MM. Order Chiroptera. Bats. 1999. In: Ducksworth W, Salter RE, Khounboline, editors. Wildlife in Lao PDR: A Status Report. Vientianne: IUCN; 1999. pp. 225-235

[42] Lee BPH, Struebig MJ, Rossiter SJ, Kingston T. Increasing concern over trade in bat souvenirs from South-East Asia. Oryx. 2015;**49**(2):204-234. DOI: 10.1017/s0030605315000034

[43] Meyer M. The Night Parade of One Hundred Demons: A Field Guide to Japanese Yokai. Self-published

[44] Batchelor J. The Ainu and their Folk-Lore. London: Religious Tract Society; 1901

[45] Sinavaiana C, Enright J. The cultural significance of the flying fox in Samoa: A legendary view. In: Proceedings of an International Conservation Conference on Pacific Island Flying Foxes. Vol. 90. 1992. pp. 36-38. US Fish and Wildlife Service Biological Report

[46] Bani E. Fruit bats of Vanuatu. Proceedings of an international conservation conference on Pacific Island flying foxes. US Fish and Wildlife Service Biological Report. 1992;**90**:123-127

[47] Lavery TH, Fasi J. Buying through your teeth: Traditional currency and conservation of flying foxes *Pteropus* spp. in Solomon Islands. Oryx. 2017;**53**:505-512. DOI: 10.1017/S0030605317001004

[48] Wendt A. Tatauing the post-colonial body. e-Spania. 1992;**42-43**:15-29

[49] Mead HM, Grove N. Ngā Pēpeha a Ngā Tīpuna: The Sayings of the Ancestors. Wellington: Victoria University Press; 2003

[50] Anonymous. A bat has Won New Zealand's Bird of the Year 2021 Competition. BBC. 2021. https://www.bbc.co.uk/newsround/59120252 [Accessed: 10/12/2021]

[51] Benson EP. Bats in south American iconography. Andean Past. 1987;**1**:165-190

[52] Benson EP. Bats in south American folklore and ancient art. Bat Conservation International. 2021;**9**(1):1-6

[53] Thompson JES. Maya hieroglyphs of the bat as Metaphorgrams. Man. 1966;**1**(2):176-184

[54] Wilbert J. Yupa Folktales. Los Angeles: UCLA Latin American Center Publications; 1974

[55] Wilbert J, Simoneau K. Folk Literature of the Mataco Indians. Los Angeles: UCLA Latin American Center Publications; 1982

[56] Lévi-Strauss C. The Raw and the Cooked. New York: Harber & Row; 1969

[57] Wagner E. The Prehistory and Ethnohistory of the Carache Area in Western Venezuela. New Haven, Connecticut: Yale University

Publications in Anthropology; 1967. p. 71

[58] Alegria RE. Ball Courts and Ceremonial Plazas in the West Indies. New Haven, Connecticut: Yale University Publications in Anthropology; 1983. 79 pp

[59] Atkinson L-G. Taino Influence on Jamaican Folk Traditions. 2010. Taino Day presentation in May

[60] Wilbert J. Folk Literature of the Warao Indians: Narrative Material and Motif Content. Los Angeles: Latin American Studies, University of California; 1970. p. 15

[61] Roth WE. An Inquiry into the Animism and Folk-Lore of the Guiana Indians. Vol. 30. Annual Report of the Bureau of American Ethnology 1908-1909; 1915. pp. 103-386

[62] Lévi-Strauss C. From Honey to Ashes. Chicago: University of Chicago Press; 1983. p. 382

[63] Guthrie J. Bats are for life, not just for Halloween. In: Financial Times. 2021. Available from: https://www.ft.com/content/ec/eecf8-ab03-4e87-918c-e8c19a7c134a [Accessed: 10/12/2021]

[64] Shedd W. Owls aren't Wise & Bats aren't Blind. New York: Harmony Books; 2000. 322 pp

[65] De Serres GF, Dallaire F, Cote M, et al. Bat rabies in the United States and Canada from 1950 through 2007 human cases with and without bat contact. Clinical Infectious Diseases. 2008;**46**: 1329-1337

[66] Hoyt JR, Kilpatrick AM, Langwig KE. Ecology and impacts of white-nose syndrome on bats. Nature Reviews Microbiology. 2021;**19**(3):196-210. DOI: 10.1038/s41579-020-00493-5

[67] Bat Conservation International. White-Nose Syndrome. Finding Solutions to Protect Bats from a Deadly Disease. 2021. https://www.batcon.org/.../white-nose-syndrome [Accessed: 10/12/2021]

[68] Tatai E. An iconographical approach to representations of the devil in medieval Hungary. Christian Demonology and Popular Mythology. 2006;**2**:54

[69] Rocha R, Aziz SA, Brook CF, Carvalho WD, Cooper-Bohannon R, Frick WF, et al. Bat conservation and zoonotic disease risk: A research agenda to prevent misguided persecution in the aftermath of COVID-19. Animal Conservation. 2020;**24**(3):303-307. DOI: 10.1111/acv.12636

[70] Schutt B. Dark Banquet: Blood and Curious Lives of Blood-Feeding Creatures. New York: Three River Press; 2008

[71] Aziz SA, Olival KJ, Bumrungsri S, Richards GC, Racey PA. The conflict between Pteropodid bats and fruit growers: Species, legislation and mitigation. In: Voigt C, Kingston T, editors. Bats in the Anthropocene: Conservation of Bats in a Changing World. Cham, Switzerland: Springer; 2016. pp. 377-426

[72] Epstein JH, Olival KJ, Pulliam JRC, Smith C, Westrum J, Hughes T, et al. *Pteropus vampyrus*, a hunted migratory species with a multinational home-range and a need for regional management. Journal of Applied Ecology. 2009;**46**:991-1002. DOI: 10.1111/j.1365-2664.2009.01699.x

[73] Dovih P. Bat hunts and disease outbreaks: Traditional bat hunting in Nagaland. Economic and Political Weekly 2015;50. Online URL: https://www.epw.in/

[74] Pennisi LA, Holland SM, Stein TV. Journal of Ecotourism. 2004;**3**(3):195-207. DOI: 10.1080/14664200508668432

[75] Tanalgo KC, Hughes AC. The potential of bat-watching tourism in raising public awareness towards bat conservation in the Philippines. Environmental Challenges. 2021;**4**:1-9. DOI: 10.1016/j.envc.2021.100140

[76] Carpenter E-S, Gomez R, Waldien DL, Sherwin RE. Photographic estimation of roosting density of Geoffroy's Rousette fruit bat *Rousettus amplexicaudatus* (Chiroptera: Pteropodidae) at Monfort bat cave, Philippines. Journal of Threatened Taxa. 2014;**6**(6):5838-5844. DOI: 10.11609/JoTT.o3522.5838-44

Section 2

Pesticide Exposure

Chapter 2

Pesticide Exposure Risks to Chiropteran Species and the Impacts on Emerging Zoonotic Diseases

Sarah Hooper and Sybill Amelon

Abstract

Neonicotinoids have been in the spotlight in the pollinator community as they persist in the soil, have high water solubility, and have been associated with negative health implications on insect pollinators. The risk of new novel pesticides, including neonicotinoids, to bats are largely unknown. Bats have unique physiology as they are the only mammals capable of true and sustained flight, and have physiological adaptations including echolocation and torpor which under current protocols for acute and chronic toxicity studies in birds and terrestrial animals are not assessed. Due to these characteristics, some have argued that bats may serve as important bioindicators for ecosystem health and pesticide use. This chapter will focus on pesticides, and discuss the increased risk of exposure, morbidity, and mortality of bats species due to their unique physiology and natural life history. Special emphasis will be on potential increased risk of zoonotic disease transmission in bats exposed to emerging contaminants that suppress their immune system or cause increased biological stress.

Keywords: bats, chiroptera, contaminants, ecotoxicology, insectivorous bats, frugivorous bats, mammals, pesticides, pesticide exposure, toxicology, viral diseases of bats, zoonotic diseases

1. Introduction

Pesticides refer to a broad range of chemicals designed to control target organisms such as insects (insecticides), plants (herbicides), and other organisms (e.g. fungicides and algicides). While judicious use of pesticides can be beneficial, there is always a risk that non-target organisms will be affected. For some compounds, the widespread negative side-effects on non-target organisms outweigh the benefits of use. The classic examples of such compounds are DDT and other organochlorides. In the early 1960s, Rachel Carson published the book *Silent Spring*, which captured the public's attention on the widespread detrimental effects that organochlorides caused in the avian population such as embryotoxicity, eggshell thinning, and severe population declines [1, 2]. As a result of her work, DDT and other organochlorides were banned in the US beginning in the early 1970s with other developed countries following suit [1, 3].

Since the banning of organochlorides, new compounds have arisen to take their place such as neonicotinoids—the most widely used insecticides in the world with over 25% market share [4, 5]. Neonicotinoids have been in the spotlight in the pollinator community as they persist in the soil, have high water solubility, and have been associated with negative health implications on pollinators [5, 6]. The risk of these newer pesticides, including neonicotinoids, to bats are largely unknown, because required pesticide risk assessment schemes for birds and mammals have been deemed inadequate to assess the risk to bats [7].

This chapter will explore the how the natural life history of bats and their physiologic adaptations (e.g. echolocation, torpor and true, sustained flight [8]) increases their risk of oral, dermal, and inhalation exposure to pesticides. Understanding how Chiropteran ecology and physiology contribute to an increased risk of pesticide exposure is critical for several reasons. 1) Bats suffering from anthropogenic stressors such as environmental pollutants are at increased risk of immunosuppression—meaning an increase in the production and shedding of viruses [9, 10]. 2) Bats serve as important bioindicators for pesticide use and ecosystem health [11]—meaning the utilization of bats as bioindicators may be key to predicting emerging and re-emerging zoonotic diseases as recent zoonotic disease outbreaks have been closely interlinked to ecosystem health [12].

2. Worldwide pesticide use and registration of pesticides

Worldwide, 3.5 million tons of pesticides were applied to farmland in 2020 [13, 14]. This is likely an underestimate, as the data was obtained from a selection of different databases—some of which are limited in their scope. For instance, one database used, the United States Geological Service (USGS) Pesticide National Synthesis Project, is a database that tracks agricultural pesticide use in the United States. In 2015, this database discontinued recording pesticides used in seed treatment applications [15]. The primary use of neonicotinoids are seed treatments [5], therefore neonicotinoids use is greatly underestimated in the United States, the second largest consumer of pesticides per the Food and Agriculture Organization of the United Nations (FAO) [16]. Additionally, many databases exclude non-agricultural pesticide products used by consumers (i.e. for home use and lawn/garden use), therefore true worldwide pesticide use is difficult to estimate.

Before a pesticide can be sold to consumers, each country undertakes a review process to validate the chemical is safe for nontarget species and efficacious. Typically, this process is allocated to governmental agencies (e.g., the Environmental Protection Agency (EPA) in the United States and the Europe Food Safety Authority (EFSA) in the European Union), or a government appointed board (e.g. The Pesticides and Toxic Chemicals Control Board in Saint Kitts and Nevis). While toxicology and ecotoxicological assessments vary by each country, the minimum requirements for most ecotoxicological studies must "(1) quantify the magnitude of acute mortality caused by the application, (2) determine the existence and extent of reproductive impairment in nontarget species from the application, and (3) determine the extent to which survival is influenced" [7, 17].

To fulfill these objectives, the study designs classically focus on obtaining the LD_{50}, the lethal dose for 50% of a group of experimental animals; LC_{50}, the lethal concentration in air or water that kills 50% of the experimental animals during a specific time period; and no-observed-effect-level (NOEL), the highest concentration of the substance at which no adverse effects are observed in the experimental animals. The experimental animals used to obtain the ecotoxicological data must be a species that is similar to those found in the desired application area. Typically rats (e.g. *Rattus norvegicus*) serve as the representative species for terrestrial mammals, the northern bobwhite (*Colinus virginianus*) or another quail species for an upland game bird, the mallard duck (*Anas platyrhynchos*) for waterfowl, and the house sparrow (*Passer domesticus*) or zebra finch (*Taeniopygia guttata*) for passerines [7, 17, 18]. Bats are classified as a terrestrial mammal for ecotoxicology testing purposes, and therefore are represented by rats, an r-selected species [19] without the physiological adaptations such as highly vascularized wings or echolocation.

3. Chiropteran life history, ecology, and physiological adaptations

3.1 Overview of bat diversity and ecological services

The order Chiroptera contains more than 1400 species—equating to over 1 in every 5 mammals is a bat [20]. Chiropteran species were previously divided into two suborders, Megachiroptera and Microchiroptera based upon morphological and behavior data. The suborder Megachiroptera consists of approximately 166 species, all belonging to the family Pteropodidae, whose diet consist of fruit, nectar, or pollen [20]. Seventeen other families compose the suborder Microchiroptera, the majority of which consume insects [20]. These suborders are still routinely used by some scientists and biologists. However, new molecular biology findings combined with morphological and behavior data have led researchers to propose two new suborders of bats, Yinpterochiroptera and Yangochiroptera, which do not coincide with the previous subordinal classification [21].

Chiroptera are critical to ecosystem health as they provide critical ecosystem services such as arthropod suppression, seed dispersal, and pollination on every continent except Antarctica [20, 22]. However, some ecosystems are at risk of degradation due to more than a quarter of the world's bat species facing the threat of extinction [23]. Additionally, over half of the species are facing population declines or have unknown population statuses further jeopardizing ecosystem health [23, 24].

Bat population declines are frequently attributed to habitat loss, White-nose Syndrome, wind energy, and more recently climate change [23, 25]. Pesticide exposure and pollutants are less recognized for causing population declines due to the lack of Chiropteran specific ecotoxicological studies. The few studies that are published directly link population declines to dermal contact and oral ingestion of pesticides [26–30].

3.2 Foraging ecology and diet

Nearly all bat species are nocturnal, with dusk and a few hours before dawn being their most active foraging times. Foraging strategies are diverse and are dependent upon the diet and the individual species morphological characteristics. The feeding behavior of bats who consume primarily fruit, frugivorous bats, are based upon their body size, dentition, and skull/jaw morphology [31]. Nectivorous bats, those whose diet consumes of primarily nectar, select flowers based upon their digestive capacity, lingual (tongue) morphology, and possibly metabolic rate [32, 33]. Relatively little is known about fishing bats, or piscivorous bats, who consume fish prey. It has been suggested that skull morphology, body size, wing morphology, and acoustic properties dictate piscivorous bats dietary niche partitioning [34, 35]. There is a lack of information on dietary morphological adaptations for the 3 sanguinivorous bat species who are obligate blood feeders.

Most is known about insectivorous bat species, those who consume insects, because most Chiropteran species fall within this category. Wing morphology, body size, and echolocation strategies influence which insect species are preyed upon [36]. For example, the aerial hawking species botta's serotine (*Eptesicus bottae*) have an average body mass of 8.9 ± 1.3 grams (adult males and nonreproductive females), broad wings with an average wingspan of 27.3 ± 1.8 mm and an average wing area of 12.6 ± 6.7 cm^2 [37, 38]. *E. bottae* fly around 5.7 meters per second and their echolocation calls allow them to detect large prey up to 21 meters away with an echo delay of one wing beat [38]. With these morphological characteristics helping to dictate their diet, DNA metabarcoding studies have revealed they consume primarily insects from the orders Auchenorrhyncha, Coleoptera, Lepidoptera, Heteroptera, and Hymenoptera [37].

By consuming agricultural pests as part of their diet, insectivorous bats are highly valued by farmers. Presently there is no worldwide estimate of the economic value of the ecological services provided by bats; however, researchers in the US have estimated that insectivorous bats provide roughly $22.9 billion dollars per year in natural pest control services to US crops [39]. Due to the high foraging activity around agricultural areas and active consumption of pests that are targeted by insecticides, nearly all of the Chiropteran ecotoxicology studies focus on insectivorous species [30].

The majority of bat species in agricultural areas actively forage over fields at night with an increase in activity at dusk or dawn due to the increased activity of crepuscular insects [40]. Due to the recognition of pesticides affecting bee health, current guidelines for pesticide applicators are to apply pesticides at dusk, dawn and late at night while pollinator activity is low to non-existent in order to avoid pollinator mortality [41].

The application of pesticides at the height of bat activity, directly increase the risk of inhalation exposure of the droplets and vapors created during pesticide applications [7]. By flying through the spray clouds when foraging over agricultural fields or the vapors along the field margins, there is an increased risk of dermal exposure as well as an increased risk of oral exposure through the consumption of prey coated during the application process. Additionally, the lights used during dusk or night application may increase bat activity of non-light aversive species due to the attraction of insects to the lights [42].

It is plausible that frugivorous and nectivorous bat species also experience a similar increase in risk to pesticide exposure especially in intensively managed orchards. Because there is only a single study documenting organochlorine pesticide in a nectivorous bat species, the lesser long-nosed bat (*Leptonycteris sanborni*), [43] and no studies, other than manipulative experimental exposure studies, on frugivorous bats [30] it is difficult to assess their exposure risk. Whereas there are numerous records of organochlorine, organophosphate, and other pesticide residues found in deceased insectivorous species indicating that pesticide exposure is occurring and can be linked to mortality [30].

3.3 Roosting ecology

Due to recent estimates of the economic value insectivorous bats (e.g. see [39]), the agricultural industry and general public have begun recognizing the valuable natural pest control services provided by insectivorous bats. This recognition has resulted in some farmers attempting to attract bats by placing bat houses near fields or leaving untouched treelines with adequate roosting structures such as tree cavities or flaking bark [7, 44, 45].

Depending upon the species, some bats may also use farmland buildings such as barns or stables, or even roost on farm homes made of brick that contain shallow cracks [46] or under shutters based upon the authors' personal observations. With roost locations located within or adjacent to crops where pesticides are applied, bats within these roosts are at increased risk of inhalation and dermal exposure from pesticide drift [47] especially those roosting on exterior walls of buildings.

While bats are within their day roosts, they exhibit a variety of activities (e.g. parturition, nursing, torpor, and grooming) depending upon the sex and season. All these activities increase the risk of exposure. For instance, grooming could result in the increased risk of oral exposure if the roost is contaminated through pesticide drift or if bats where in direct contact with the pesticides during the application process. These roosts tend to be continuously habited for weeks to months and many are reutilized each year [48] leading to the potential for continuous long-term exposure if the roosts are contaminated. Bats also utilize night roosts in order to conserve energy, avoid predators, digest consumed food, to exchange information and for social interactions [49]. These night roosts have similar increased risks if they are in areas exposed to pesticide drift.

Some species select caves to roost for hibernation or may utilize them throughout the year. By crawling on cave walls and drinking from underground water sources, they are at risk of dermal and oral exposure because cave sediments, groundwater, and drip water samples in surveyed areas have all been reported to contain a variety of organochlorines and other pesticide metabolites [50–53].

3.4 Drinking

Bats are unable to obtain enough water through their diet, and so must rely upon drinking to meet daily needs. In addition to drinking potentially contaminated karst water, bats also drink from tailing ponds which can be high in cyanide, heavy metals, and occasionally pesticides depending upon the location of the mining operation [53]. Within the agricultural setting, bats have been observed to drink from irrigation canals, small ponds, and water troughs with the latter two often containing pesticides from overland flow [53] and thereby increasing the risk of oral exposure.

3.5 Reproduction

It is generally believed that bats reproduce once per year with most species giving birth to 1 pup or occasionally twins (e.g. big brown bat, *Eptesicus fuscus*) [54]. The exception to this is bats belonging to the genus Lasiurus who can give birth to up to four pups. For more detailed information on the reproductive biology of bats see [54].

Actual pesticide exposure risks to pups is difficult to assess. Two older studies assessing banned organochlorine pesticides such as DDE documented maternal transfer and resulted in offspring mortality [55, 56]. There is only a single study assessing reproductive effects for a commercially available fungicide, tebuconazole. This study coated papaya with the manufacturer's recommended application concentration of tebuconazole. When great fruit-eating bats (*Artibeus lituratus*) consumed the fruit, testicular and epididymal histomorphological changes were observed, suggesting that the fungicide tebuconazole may impact reproductive capacities in males [57]. Unfortunately, reproductive females were not evaluated.

Despite the lack of studies on contaminants in actively reproducing female bats and/or their offspring, it is probable that pups can be exposed in utero based upon numerous studies in humans and other mammals documenting placental transfer of commonly sold pesticides including neonicotinoids [58–60]. Furthermore, because pesticides and their metabolites are often detected in mammalian milk [61–63], pups are likely exposed to pesticides for approximately 3–4 weeks after birth while nursing.

4. Chiropteran physiology

4.1 Skeletal morphology and flight

Unique skeletal system and integumentary system adaptations combined make bats the only mammals capable of true, sustained flight. Chiropteran species have a strong axial skeleton with non-pneumatized, marrow-filled long bones and elongated metacarpal bones and phalanges [64]. A bilayered 1–2 cell thick epidermis (e.g. <s10 μm in the epauletted fruit bat, *Epomophorus wahlbergi*) separated by a non-distinct dermis or hypodermis form the extremely thin wing membrane. Within the non-distinct dermis there are highly vascularized connective tissues [65]. Collagen and elastic fibers within the connective tissues serve as "fingerprints" that can be used to identify individuals [66] and are responsible for providing the mechanical strength needed for flight [65, 66].

The wings may serve additional purposes other than flight. The thin, and highly vascularized wing tissue has been suggested to allow diffusion of oxygen and carbon dioxide between the environment and the vasculature, thereby contributing to a bat's total gas exchange [65]. Wing membranes are essential to thermoregulation [67] and also water exchange [8, 65]. These adaptations allowing water and gas exchange through the wing membrane could also increase the risk of dermal absorption of pesticides—especially considering the wing membrane of bats accounts for up to 85% of a bat's total body surface [65].

When bats fly, it is energetically expensive, especially since bats flap continuously rather than glide [36]. To meet this energy demand, food intake is high. When foraging each night, frugivores consume up to 2.5 times their body weight in fruit [68], insectivores consume up to 1.3 times their body weight in insects [7], and some nectivorous species consume nectar from 80 to 100 flowers each night [69]. With high food consumption, there is an increased risk of oral exposure to all bat species.

4.2 Torpor

Torpor or heterothermic responses are "facultative, reversible decreases in metabolic rate and body temperature in response to environmental cues" [70]. All species of bats may select to employ daily torpor, the lowering of their body temperature set point, respiratory rate, and metabolic rate for a single circadian cycle or a portion of a circadian cycle [49, 71] to help offset the high metabolic and thermoregulatory costs [71]. Hibernating species may engage in multi-day torpor bouts which can save them up to 99% of their daily energy requirements [72]—allowing them to survive winter when food availability is reduced [71].

Throughout hibernation (multi-day torpor), bats use white adipose tissue as their primary fuel source [73]. When arousing from torpor, especially multi-day torpor bouts, they may engage in passive rewarming which can reduce arousal costs by 50% or more [74]. Even with passive rewarming, torpor arousal is still energetically demanding and bats must mobilize a portion of their brown adipose tissue to generate heat [73].

During the fall, when hibernating bats are developing large fat stores, the risk of oral exposure to pesticides increases as they consume a greater amount of prey. Concurrently, lipophilic (fat soluble) pesticide residues are stored within the adipose tissue (see Table 1 in [30] for studies which measured pesticides in Chiropteran adipose tissue) as a large portion of pesticides, especially within the insecticide class, are lipophilic [75]. Hibernating bats may be at risk of continuous lipophilic pesticide exposure as bats mobilize their white adipose tissue. During arousal the rapid mobilization of brown adipose tissue may cause an increased risk of a sudden large dose of lipophilic pesticides to be released.

4.3 Echolocation

It is a common myth that bats are blind, as in fact they have color vision and may be sensitive to UV light [76]. The majority of species within the Pteropodidae family (Old World Fruit bats such as flying foxes), use their acute vision to perceive their environment and forage for food rather than rely upon echolocation. Whereas bats in the other families are believed to use echolocation as the primary way to forage and navigate through their environments [77]. When these species echolocate, they produce high frequency sounds or tongue clicks [78]. Their auditory system processes the echoes and enables the 3D perception of the object and the spatial location of the object. For more detailed information on the adaptive sonar behavior of bats and echolocation auditory mechanisms see [77, 79].

Echolocation has been suggested to contribute to the global success of a wide range of bat species [78]. While echolocation may not increase a bats risk to pesticide exposure, it can be negatively affected by sublethal concentrations of pesticides [80].

4.4 Other considerations for reference species

At sublethal doses, the neonicotinoid imidacloprid suppresses the expression of genes related to echolocation (e.g. FOXP2) [80], interferes with the spatial memory [81], and leads to inflammation and mitochondria dysfunction-related apoptosis [80]. This type of ecotoxicological data suggests that using mammals who do not have echolocation capabilities may need to be reassessed as a reference species for bats.

Other considerations for reference species selection should consider that other mammals do not have the physiological adaptations described above. Additionally, rats, the most commonly used reference species, have a relatively short lifespan of 3 years or less in a laboratory or pet setting, with the average lifespan of a wild rat being less than 1 year [82]. Free-ranging bats live much longer with some species living for at least 32 years based upon recapturing banded individuals [33]. The longer lifespans of bats put them at increased risk of bioaccumulation of pesticides and other contaminants such as heavy metals.

5. Zoonotic disease risks

It is estimated that over 60% of newly emerging diseases are zoonotic with approximately 3 out of every 4 diseases being the result of spillover from wildlife [84, 85]. Due to the ongoing worldwide pandemic COVID-19, caused by the severe acute respiratory syndrome coronavirus 2 (SARS-CoV-2), bats are often cited by the media and some scientists as the source of this zoonotic disease as well as other zoonotic viral diseases such as Ebola [86]. COVID-19 is an emerging infectious disease but at the time of this writing there is no evidence it is a zoonotic disease [87].

A zoonotic disease, or zoonosis, is defined by the World Health Organization as "any disease or infection that is naturally transmissible from vertebrate animals to humans [88]". While the origin of SARS-CoV-2 remains unknown [89], there is no evidence that bats serve as the reservoir of the virus. This is because a reservoir is "a population, species or community assemblage of different species in a given geographic area in which a microorganism naturally occurs and is indefinitely maintained" [86]. Since SARS-CoV-2 has not been detected in any bat species, they cannot serve as reservoirs of the virus [86]. The miscommunication seems to originate from the misunderstanding of a report of a newly identified coronavirus circulating within intermediate horseshoe bats (*Rhinolophus affinis*) which exhibits

96% similarity to SARS-CoV-2 virus [90]. This newly identified coronavirus is suggested to be the ancestral origin of the SARS-CoV-2 virus [90], but it is not the same virus that is causing the COVID-19 pandemic. For a complete review on effective communication regarding zoonotic diseases and the terminology commonly misunderstood and miscommunicated in relation to bats see [86].

While COVID-19 may not be a zoonotic virus, bats do serve as the reservoir for other zoonotic viruses such as rabies and Nipah virus [86]. When new zoonotic diseases arise, they are most often the result of a spillover event from bats or other wildlife reservoir species to humans or another mammalian host [10, 91].

Spillover events are complex processes and are rare [9]. When spillover events do occur, there must be a "perfect storm" or a situation where several factors need to be present for the spillover event to occur. Two of those factors, pathogen shedding and environmental conditions, are discussed below. For an extensively detailed description of all potential factors see Plowright et al. 2015 [92].

Before describing these factors, it must be noted that the immune system of bats is not well understood. The limited research conducted on Chiropteran immune systems verify there is much diversity among species (e.g. the interferon gene locus is contracted in the black flying fox (*Pteropus alecto*) whereas expanded in the Egyptian fruit bat (*Rousettus aegyptiacus*) [93–95]). While this diversity exists, their immune systems uniformly appear to allow chronic/latent viral infections where the host remains asymptomatic and viral shedding does not occur. This benign virus-host relationship is more commonly recognized as the Susceptible-Infectious-Latent-Infectious (SILI) hypothesis [10].

The SILI hypothesis suggests that when bats are stressed by physiological and/or environmental stressors, the virus(es) can reactivate due to a suppression of the host's immune system [10]. These viruses may remain nonpathogenic within bats, but the reactivation of the viruses results in viral shedding [10]. Viral shedding leads to an increased risk of spillover events as these viruses may cause severe disease in other mammals [86, 96].

Anthropogenic changes such as urbanization and high human population density are known physiological and environmental stressors to bats due to the loss, degradation and/or fragmentation of suitable habitats [97]. While these anthropogenic stressors are well studied, there are few experimental studies assessing if pesticides and other pollutants are sublethal stressors as most Chiropteran contaminant studies are observational.

Pesticides are classified as stressors in other mammalian species as they are known to cause oxidative stress, genotoxic effects, and suppress the immune system [98–101]. Two studies conducted in the fruit bat *A. lituratus* measured oxidative stress when exposed to fruit sprayed with the organochlorine insecticide endosulfan and the pyrethroid insecticide deltamethrin. The metabolism of the pesticides increased the generation of reactive oxygen species (ROS) including nitric oxide and hydrogen peroxide leading to liver morphological changes [102] and oxidative damage to both liver and the pectoral muscle [102, 103]. Since fruit bats are believed to be highly resistant to oxidative stress in part due to the high level of antioxidants in their diet [104], this finding is significant. Especially since it is well established that chronically increased oxidative stress leads to a variety of pathophysiological conditions [105] and directly alters the immune system [106]. It is unknown if increased oxidative stress leads to increased viral shedding, thereby increasing the risk of zoonotic disease and should be evaluated in future studies.

Oxidative damage has been shown to induce genotoxicity when humans have been exposed to some contaminants such as lead and toluene [107, 108]. Two commonly employed genotoxicity tests are the micronucleus test and the comet assay. The micronucleus test assesses chromosomal damage (see [109] for detailed review of the technique) while the comet assay detects a wide array of DNA damage such as DNA strand breaks (single and double stranded) and incomplete excision repair (see [110] for detailed review of the technique).

Micronucleus tests conducted in the insectivorous bat species, Parnell's mustached bat (*Pteronotus mexicanus*), documented that *P. mexicanus* roosting in caves surrounded by agriculture and foraging over pesticide treated fields, had a higher number of micronuclei than bats roosting and foraging in areas where no more than 15% of the surrounding land-use was agriculture [111]. Another study assessing micronuclei and other nuclear abnormalities in insectivorous, frugivorous, and nectivorous bats discovered that nectivorous bats did not have any significant nuclear changes in agricultural or urban areas [112]. Frugivorous and insectivorous bats however did have significantly higher micronuclei in urban and agricultural areas with other nuclear changes such as binucleated cells being commonly observed in urban areas [112]. A third study assessing genotoxicity using the comet assay documented that banana bats, *Neoromicia nana*, foraging over South African waste water treatment plants containing pesticides and other pollutants had significantly higher DNA damage compared to bats foraging at reference sites within Umdoni Park [113]. All three studies are limited from the perspective that none measured actual pesticide concentrations in tissues or bodily fluids (e.g. urine). Although it cannot be solely concluded that the genotoxic effects were truly from pesticide exposure, research in other mammalian species documents a clear link between pesticide exposure and induced genotoxicity [99]. This fact identifies along with the three Chiropteran studies suggesting pesticide exposure can induce genotoxic changes, highlights a large research need for assessing if pesticides are genotoxic to bats and if it leads to viral shedding via immunosuppression mechanisms.

Although there are few studies measuring the ecotoxicological effects of pesticides on bat species, the available studies show that some species suffer oxidative damage, genotoxic changes, and mortality. This suggests that compounding anthropogenic stressors, including environments with heavy pesticide use, may cause environmental conditions to become favorable for a spill-over event to occur by causing physiological stress in insectivorous and frugivorous bats. This compounded physiological stress may result in viral reactivation and viral shedding by suppressing bats' unique immune system. Because bats sustain a high viral load and are phylogenetically close to humans, the risk of zoonotic disease transmission from spill-over events is greatest from bats compared to other species [114].

6. Conclusions

Ecotoxicological studies in wildlife species use rats as the reference species for terrestrial wildlife, including bats. With over 1400 bat species, and unique morphological characteristics such as highly vascularized wing membranes composing up to 85% of their body surface and unique physiology such as echolocation [20, 59], species lacking these characteristics (e.g. rats) may not be adequate when assessing effects of pesticide exposure.

Bats natural life history combined with these unique adaptations leads to the increased risk of pesticide exposure. Due to their high risk of exposure, bats can serve as important bioindicators for both pesticide use and ecosystem health [11]. Early detection of zoonotic diseases may be possible by monitoring ecosystem health through bioindicators such as bats [11, 12], because unhealthy ecosystems exert physiological and environmental stressors. These stressors, including pesticides, may lead to a spillover event where a wildlife species immune system becomes suppressed and pathogen shedding becomes possible [9, 10, 91, 92].

Conflict of interest

The authors declare no conflict of interest.

Author details

Sarah Hooper[1]* and Sybill Amelon[2]

1 Ross University School of Veterinary Medicine, Basseterre, Saint Kitts

2 USDA Forest Service Northern Research Stations, Columbia, USA

*Address all correspondence to: shooper@rossvet.edu.kn

References

[1] Pollock CG. Silent spring revisited: A 21st-century look at the effect of pesticides on wildlife. Journal of Avian Medicine and Surgery. 2001;**15**(1): 50-53

[2] Carson R. Silent spring 40th anniversary edition. Boston, MA: Houghton Mifflin; 2002. p. 400

[3] United States Environmental Protection Agency. DDT -A Brief History and Status Available from: https://www.epa.gov/ingredients-used-pesticide-products/ddt-brief-history-and-status [Accessed: 05-09-2021]

[4] Simon-Delso N, Amaral-Rogers V, Belzunces LP, Bonmatin JM, Chagnon M, Downs C, et al. Systemic insecticides (neonicotinoids and fipronil): Trends, uses, mode of action and metabolites. Environmental Science and Pollution Research International. 2015;**22**(1):5-34. DOI: 10.1007/s11356-014-3470-y

[5] Grout TA, Koenig PA, Kapuvari JK, SH MA. Neonicotinoid Insecticides in New York State. Economic Benefits and Risk to Pollinators. Ithaca, NY: Cornell University; 2020. p. 432

[6] Lu C, Hung Y-T, Cheng Q. A review of sub-lethal neonicotinoid insecticides exposure and effects on pollinators. Current Pollution Reports. 2020;**6**(2):137-151. DOI: 10.1007/s40726-020-00142-8

[7] EFSA Panel on Plant Protection Products and their Residues (PPR), Hernández-Jerez A, Adriaanse P, Aldrich A, Berny P, et al. Scientific statement on the coverage of bats by the current pesticide risk assessment for birds and mammals. EFSA Journal. 2019;**17**(7):e05758. DOI: 10.2903/j.efsa.2019.5758

[8] Kunz TH, Parsons S. Ecological and behavioral methods for the study of bats. 2nd ed. Baltimore: Johns Hopkins University Press; 2009. p. 920

[9] Subudhi S, Rapin N, Misra V. Immune system modulation and viral persistence in bats: Understanding viral spillover. Viruses. 2019;**11**(2):192. DOI: 10.3390/v11020192

[10] Plowright RK, Peel AJ, Streicker DG, Gilbert AT, McCallum H, Wood J, et al. Transmission or within-host dynamics driving pulses of zoonotic viruses in reservoir–host populations. PLoS Neglected Tropical Diseases. 2016;**10**(8):e0004796. DOI: 10.1371/journal.pntd.0004796

[11] Jones G, Jacobs DS, Kunz TH, Willig MR, Racey PA. Carpe noctem: The importance of bats as bioindicators. Endangered Species Research. 2009; **8**(1-2):93-115. DOI: 10.3354/esr00182

[12] United Nations Environmental Program. UNEP Frontiers 2016 Report: Emerging Issues of Environmental Concern. Nairobi: United Nations Environment Programme; 2016. p. 76

[13] Sharma A, Kumar V, Shahzad B, Tanveer M, Sidhu GPS, Handa N, et al. Worldwide pesticide usage and its impacts on ecosystem. SN Applied Sciences. 2019;**1**(11):1446. DOI: 10.1007/s42452-019-1485-1

[14] Sharma A, Shukla A, Attri K, Kumar M, Kumar P, Suttee A, et al. Global trends in pesticides: A looming threat and viable alternatives. Ecotoxicology and Environmental Safety. 2020;**201**:110812. DOI: 10.1016/j.ecoenv.2020.110812

[15] United States Geological Service. National Water-Quality Assessment (NAWQA) Project Pesticide National Synthesis Project Available from: https://water.usgs.gov/nawqa/pnsp/usage/maps/ [Accessed: 24-08-2021]

[16] Food and Agriculture Organization of the United Nations. FAOSTAT Pesticide Use 2019 Available from: http://www.fao.org/faostat/en/#data/RP/visualize [Accessed: 01-09-2021]

[17] United States Environmental Protection Agency. Ecological Effects Test Guidelines OCSPP 850.2500: Field Testing for Terrestrial Wildlife. Washington, DC: EPA Office of Chemical Safety and Pollution Prevention; 2012.

[18] United States Environmental Protection Agency. Ecological Effects Test Guidelines OCSPP 850.2100: Avian Acute Oral Toxicity Test. Washington, DC: EPA Office of Chemical Safety and Pollution Prevention; 2012.

[19] MacArthur RH, Wilson EO. The Theory of Island Biogeography. Princeton, New Jersey: Princeton University Press; 2001. p. 224

[20] Simmons N. Bat Species of the World: A Taxonomic and Geographic Database 2020 Available from: https://batnames.org [Accessed: 01-09-2021]

[21] Lei M, Dong D. Phylogenomic analyses of bat subordinal relationships based on transcriptome data. Scientific Reports. 2016;**6**(1):27726. DOI: 10.1038/srep27726

[22] Kunz TH, Braun de Torrez E, Bauer D, Lobova T, Fleming TH. Ecosystem services provided by bats. Annals of the New York Academy of Sciences. 2011;**1223**:1-38. DOI: 10.1111/j.1749-6632.2011.06004.x

[23] Frick WF, Kingston T, Flanders J. A review of the major threats and challenges to global bat conservation. Annals of the New York Academy of Sciences. 2020;**1469**(1):5-25. DOI: 10.1111/nyas.14045

[24] Ghanem SJ, Voigt CC. Increasing awareness of ecosystem services provided by bats. Advances in the Study of Behavior. 2012;**44**:279-302. DOI: 10.1016/b978-0-12-394288-3.00007-1

[25] O'Shea TJ, Cryan PM, Hayman DTS, Plowright RK, Streicker DG. Multiple mortality events in bats: A global review. Mammal Review. 2016;**46**(3):175-190. DOI: 10.1111/mam.12064

[26] Clark DR Jr. DDT and the decline of free-tailed bats (Tadarida brasiliensis) at Carlsbad Cavern, New Mexico. Archives of Environmental Contamination and Toxicology. 2001;**40**(4):537-543. DOI: 10.1007/s002440010207

[27] Clark DR. How sensitive are bats to insecticides? Wildlife Society Bulletin (1973-2006). 1988;**16**(4):399-403

[28] Mitchell-Jones AJ, Cooke AS, Boyd IL, Stebbings RE. Bats and remedial timber treatment chemicals a review. Mammal Review. 1989;**19**(3):93-110. DOI: 10.1111/j.1365-2907.1989.tb00405.x

[29] Schmidt A, Brack V, Rommé R, Tyrell K, Gehrt A. Bioaccumulation of pesticides in bats from Missouri. In: Pesticides and Wildlife. ACS Symposium Series. 771. American Chemical Society; 2000. pp. 8-21. DOI: 10.1021/bk-2001-0771.ch002

[30] Torquetti CG, Guimaraes ATB, Soto-Blanco B. Exposure to pesticides in bats. Science of the Total Environment. 2021;**755**(Pt 1):142509. DOI: 10.1016/j.scitotenv.2020.142509

[31] Dumont ER, O'Neal R. Food hardness and feeding behavior in old world fruit bats (Pteropodidae). Journal of Mammalogy. 2004;**85**(1):8-14. DOI: 10.1644/BOS-107

[32] Castillo-Figueroa D, Pérez-Torres J. On the development of a trait-based approach for studying Neotropical bats. Papéis Avulsos de Zoologia. 2021;**61**. DOI: 10.11606/1807-0205/2021.61.24

[33] Ayala-Berdon J, Galicia R, Flores-Ortiz C, Medellin RA, Schondube JE. Digestive capacities allow the Mexican long-nosed bat (Leptonycteris nivalis) to live in cold environments. Comparative Biochemistry and Physiology. Part A, Molecular & Integrative Physiology. 2013;**164**(4):622-628. DOI: 10.1016/j.cbpa.2013.01.015

[34] Santana SE, Cheung E. Go big or go fish: Morphological specializations in carnivorous bats. Proceedings of the Royal Society B: Biological Sciences. 2016;**283**(1830):20160615. DOI: 10.1098/rspb.2016.0615

[35] Aizpurua O, Alberdi A. Ecology and evolutionary biology of fishing bats. Mammal Review. 2018;**48**(4):284-297. DOI: 10.1111/mam.12136

[36] Norberg UM, Rayner JM. Ecological morphology and flight in bats (Mammalia; Chiroptera): Wing adaptations, flight performance, foraging strategy and echolocation. Philosophical Transactions of the Royal Society of London B, Biological Sciences. 1987;**316**(1179):335-427. DOI: 10.1098/RSTB.1987.0030

[37] Korine C. Botta's Serotine Eptesicus bottae (Peters, 1869). In: Hackländer K, Zachos FE, editors. Handbook of the Mammals of Europe. Cham: Springer International Publishing; 2020. pp. 1-9

[38] Holderied MW, Korine C, Fenton MB, Parsons S, Robson S, Jones G. Echolocation call intensity in the aerial hawking bat Eptesicus bottae (Vespertilionidae) studied using stereo videogrammetry. Journal of Experimental Biology. 2005;**208**(7): 1321-1327. DOI: 10.1242/jeb.01528

[39] Boyles JG, Cryan PM, McCracken GF, Kunz TH. Economic importance of bats in agriculture. Science. 2011;**332**(6025):41-42. DOI: 10.1126/science.1201366

[40] Jones G, Rydell J. Foraging strategy and predation risk as factors influencing emergence time in echolocating bats. Philosophical Transactions of the Royal Society of London. Series B, Biological Sciences. 1994;**346**(1318):445-455. DOI: 10.1098/rstb.1994.0161

[41] Management CfIP. 50 Ways to Treat Your Pesticide Pest Management Professional Edition 2021 Available from: https://pesticidestewardship.org/resources/.

[42] Rowse E, Lewanzik D, Stone E, Harris S, Jones G. Dark matters: The effects of artificial lighting on bats. In: Bats in the Anthropocene: Conservation of Bats in a Changing World. Cham: Springer; 2016. pp. 187-213

[43] Reidinger RF. Organochlorine residues in adults of six southwestern bat species. The Journal of Wildlife Management. 1976;**40**(4):677-680. DOI: 10.2307/3800563

[44] Rueegger N. Bat boxes — A review of their use and application, past, present and future. Acta Chiropterologica. 2016;**18**(1):279-299, 21. DOI 10.3161/15081109ACC2016.18.1.017

[45] Kyheröinen E-M, Aulagnier S, Dekker J, Dubourg-Savage M-J, Ferrer B, Gazaryan S, et al. Guidance on the conservation and management of critical feeding areas and commuting routes for bats. Bonn, Germany: UNEP/EUROBATS; 2019. p. 109

[46] Marnell F, Presetnik P. Protection of Overground Roosts for Bats Bonn, Germany: UNEP/EUROBATS Secretariat; 2010. p. 57

[47] Johnston JJ, Johnston JJ. Pesticides and wildlife. ACS Publications; 2001. DOI: 10.1021/bk-2001-0771

[48] Voigt CC, Phelps KL, Aguirre LF, Schoeman MC, Vanitharani J, Zubaid A. Bats and buildings: The conservation of

synanthropic bats. In: Bats in the Anthropocene: Conservation of Bats in a Changing World. Cham: Springer; 2016. pp. 427-462. DOI: 10.1007/978-3-319-25220-9_14

[49] Kunz TH. Ecology of bats. New York: Springer Science & Business Media; 2013. DOI: 10.1007/978-1-4613-3421-7

[50] Boccelli R, Del Re AAM, Paoletti MG, Cell M. Pesticides in mud, water, and bat guano in some veneto caves. In: Coleman D, Foissner W, Paoletti MG, editors. Soil Biota, Nutrient Cycling and Farming Systems. Boca Raton, Florida: Lewis publication; 1993. pp. 53-60

[51] Rodríguez AGP, López MIR, Casillas ÁD, León JAA, Banik SD. Impact of pesticides in karst groundwater. Review of recent trends in Yucatan, Mexico. Groundwater for Sustainable Development. 2018;**7**:20-29

[52] Modrá H, Gruberová E, Konečný O, Ulmann V, Kaucká P, Vlková M, et al. Influx and concentration of triazine pesticides in the Amaterska cave system, Moravian Karst, Czech Republic. Journal of Soils and Sediments. 2018;**18**(2):640-647. DOI: 10.1007/s11368-017-1831-0

[53] Korine C, Adams R, Russo D, Fisher-Phelps M, Jacobs D. Bats and water: Anthropogenic alterations threaten global bat populations. In: Voigt CC, Kingston T, editors. Bats in the Anthropocene: Conservation of Bats in a Changing World. Cham: Springer International Publishing; 2016. pp. 215-241. DOI: 10.1007/978-3-319-25220-9_8

[54] Crichton EG, Krutzsch PH. Reproductive Biology of Bats. San Diego, California: Academic Press; 2000. p. 510

[55] Clark DR, Lamont TG. Organochlorine residues and reproduction in the big brown bat. The Journal of Wildlife Management. 1976;**40**(2):249-254. DOI: 10.2307/3800422

[56] Thies M, McBee K. Cross-placental transfer of organochlorine pesticides in mexican free-tailed bats from Oklahoma and New Mexico. Archives of Environmental Contamination and Toxicology. 1994;**27**(2):239-242. DOI: 10.1007/BF00214268

[57] Machado-Neves M, Neto MJO, Miranda DC, Souza ACF, Castro MM, Sertorio MN, et al. Dietary exposure to tebuconazole affects testicular and epididymal histomorphometry in frugivorous bats. Bulletin of Environmental Contamination and Toxicology. 2018;**101**(2):197-204. DOI: 10.1007/s00128-018-2377-6

[58] Saxena MC, Siddiqui MK, Bhargava AK, Murti CR, Kutty D. Placental transfer of pesticides in humans. Archives of Toxicology. 1981;**48**(2-3):127-134. DOI: 10.1007/BF00310482

[59] Ohno S, Ikenaka Y, Onaru K, Kubo S, Sakata N, Hirano T, et al. Quantitative elucidation of maternal-to-fetal transfer of neonicotinoid pesticide clothianidin and its metabolites in mice. Toxicology Letters. 2020;**322**:32-38. DOI: 10.1016/j.toxlet.2020.01.003

[60] Mathiesen L, Buerki-Thurnherr T, Pastuschek J, Aengenheister L, Knudsen LE. Fetal exposure to environmental chemicals; insights from placental perfusion studies. Placenta. 2021;**106**:58-66. DOI: 10.1016/j.placenta.2021.01.025

[61] Chen D, Liu Z, Barrett H, Han J, Lv B, Li Y, et al. Nationwide biomonitoring of neonicotinoid insecticides in breast milk and health risk assessment to nursing infants in the chinese population. Journal of Agricultural and Food Chemistry.

2020;**68**(47):13906-13915. DOI: 10.1021/acs.jafc.0c05769

[62] Garcês A, Pires I, Rodrigues P. Teratological effects of pesticides in vertebrates: A review. Journal of Environmental Science and Health. Part. B. 2020;**55**(1):75-89. DOI: 10.1080/03601234.2019.1660562

[63] Ichikawa G, Kuribayashi R, Ikenaka Y, Ichise T, Nakayama SM, Ishizuka M, et al. LC-ESI/MS/MS analysis of neonicotinoids in urine of very low birth weight infants at birth. PLoS One. 2019;**14**(7):e0219208. DOI: 10.1371/journal.pone.0219208

[64] Gatesy SM, Middleton KM. Skeletal adaptations for flight. In: Hall BK, editor. Fins into Limbs: Evolution, Development, and Transformation. Chicago, Illinois: University of Chicago Press; 2008. pp. 269-283

[65] Makanya AN, Mortola JP. The structural design of the bat wing web and its possible role in gas exchange. Journal of Anatomy. 2007;**211**(6): 687-697. DOI: 10.1111/j.1469-7580.2007.00817.x

[66] Amelon SK, Hooper SE, Womack KM. Bat wing biometrics: Using collagen–elastin bundles in bat wings as a unique individual identifier. Journal of Mammalogy. 2017;**98**(3): 744-751. DOI: 10.1093/jmammal/gyx018

[67] Thomson S, Speakman J. Absorption of visible spectrum radiation by the wing membranes of living pteropodid bats. Journal of Comparative Physiology B. 1999;**169**(3):187-194. DOI: 10.1007/s003600050210

[68] Thomas DW. Fruit intake and energy budgets of frugivorous bats. Physiological Zoology. 1984;**57**(4): 457-467. DOI: 10.1086/physzool.57.4.30163347

[69] Horner MA, Fleming T, Sahley C. Foraging behaviour and energetics of a nectar-feeding bat, Leptonycteris curasoae (Chiroptera: Phyllostomidae). Journal of Zoology. 1998;**244**:575-586. DOI: 10.1111/J.1469-7998.1998.TB00062.X

[70] McKechnie AE, Mzilikazi N. Heterothermy in afrotropical mammals and birds: A review. Integrative and Comparative Biology. 2011;**51**(3): 349-363. DOI: 10.1093/icb/icr035

[71] Willis CK, Brigham RM. Defining torpor in free-ranging bats: Experimental evaluation of external temperature-sensitive radiotransmitters and the concept of active temperature. Journal of Comparative Physiology. B. 2003;**173**(5):379-389. DOI: 10.1007/s00360-003-0343-y

[72] Wang LC, Wolowyk MW. Torpor in mammals and birds. Canadian Journal of Zoology. 1988;**66**(1):133-137. DOI: 10.1139/z88-017

[73] Ballinger M, Andrews M. Nature's fat-burning machine: Brown adipose tissue in a hibernating mammal. The Journal of Experimental Biology. 2018;**221**:jeb162586. DOI: 10.1242/jeb.162586

[74] Currie SE, Noy K, Geiser F. Passive rewarming from torpor in hibernating bats: Minimizing metabolic costs and cardiac demands. American Journal of Physiology. Regulatory, Integrative and Comparative Physiology. 2015;**308**(1): R34-R41. DOI: 10.1152/ajpregu.00341.2014

[75] He Z, Wang Y, Wang L, Peng Y, Wang W, Liu X. Determination of 255 pesticides in edible vegetable oils using QuEChERS method and gas chromatography tandem mass spectrometry. Analytical and Bioanalytical Chemistry. 2017;**409**(4):1017-1030. DOI: 10.1007/s00216-016-0016-9

[76] Wang D, Oakley T, Mower J, Shimmin LC, Yim S, Honeycutt RL, et al. Molecular evolution of bat color vision genes. Molecular Biology and Evolution. 2004;**21**(2):295-302. DOI: 10.1093/molbev/msh015

[77] Moss CF, Surlykke A. Probing the natural scene by echolocation in bats. Frontiers in Behavioral Neuroscience. 2010;**4**:33. DOI: 10.3389/fnbeh.2010.00033

[78] Jones G, Teeling EC. The evolution of echolocation in bats. Trends in Ecology & Evolution. 2006;**21**(3): 149-156. DOI: 10.1016/j.tree.2006.01.001

[79] Moss CF. Auditory mechanisms of echolocation in bats. In: Oxford Research Encyclopedia of Neuroscience. Oxford University Press; 2018. DOI: 10.1093/acrefore/9780190264086.013.102

[80] Wu C-H, Lin C-L, Wang S-E, Lu C-W. Effects of imidacloprid, a neonicotinoid insecticide, on the echolocation system of insectivorous bats. Pesticide Biochemistry and Physiology. 2020;**163**:94-101. DOI: 10.1016/j.pestbp.2019.10.010

[81] Hsiao CJ, Lin CL, Lin TY, Wang SE, Wu CH. Imidacloprid toxicity impairs spatial memory of echolocation bats through neural apoptosis in hippocampal CA1 and medial entorhinal cortex areas. Neuroreport. 2016;**27**(6):462-468. DOI: 10.1097/WNR.0000000000000562

[82] Sengupta P. A scientific review of age determination for a laboratory rat: How old is it in comparison with human age. BioMed International. 2011;**2**(2): 81-89

[83] White JP, Nordquist GE, Kaarakka HM. Longevity records of five male little brown bats (Myotis lucifugus) in Northwest Wisconsin. Northeastern Naturalist. 2019;**26**(4). DOI: 10.1656/045.026.0401

[84] Woolhouse MEJ, Gowtage-Sequeria S. Host range and emerging and reemerging pathogens. Emerging Infectious Diseases. 2005;**11**(12):1842-1847. DOI: 10.3201/eid1112.050997

[85] Jones KE, Patel NG, Levy MA, Storeygard A, Balk D, Gittleman JL, et al. Global trends in emerging infectious diseases. Nature. 2008;**451**(7181): 990-993. DOI: 10.1038/nature06536

[86] Shapiro JT, Víquez-R L, Leopardi S, Vicente-Santos A, Mendenhall IH, Frick WF, et al. Setting the terms for zoonotic diseases: Effective communication for research, conservation, and public policy. Viruses. 2021;**13**(7):1356. DOI: 10.3390/v13071356

[87] Haider N, Rothman-Ostrow P, Osman AY, Arruda LB, Macfarlane-Berry L, Elton L, et al. COVID-19-Zoonosis or emerging infectious disease? Frontiers in Public Health. 2020;**8**:596944. DOI: 10.3389/fpubh.2020.596944

[88] World Health Organization. Zoonosis. WHO; 2020 Available from: https://www.who.int/news-room/fact-sheets/detail/zoonoses [Accessed 01-09-2021]

[89] Sills J, Bloom JD, Chan YA, Baric RS, Bjorkman PJ, Cobey S, et al. Investigate the origins of COVID-19. Science. 2021;**372**(6543):694. DOI: 10.1126/science.abj0016

[90] Andersen KG, Rambaut A, Lipkin WI, Holmes EC, Garry RF. The proximal origin of SARS-CoV-2. Nature Medicine. 2020;**26**(4):450-452. DOI: 10.1038/s41591-020-0820-9

[91] Plowright RK, Parrish CR, McCallum H, Hudson PJ, Ko AI,

Graham AL, et al. Pathways to zoonotic spillover. Nature Reviews. Microbiology. 2017;**15**(8):502-510. DOI: 10.1038/nrmicro.2017.45

[92] Plowright RK, Eby P, Hudson PJ, Smith IL, Westcott D, Bryden WL, et al. Ecological dynamics of emerging bat virus spillover. Proceedings of the Biological Sciences. 2015;**282**(1798):20142124. DOI: 10.1098/rspb.2014.2124

[93] Zhou P, Cowled C, Todd S, Crameri G, Virtue ER, Marsh GA, et al. Type III IFNs in pteropid bats: Differential expression patterns provide evidence for distinct roles in antiviral immunity. Journal of Immunology. 2011;**186**(5):3138-3147. DOI: 10.4049/jimmunol.1003115

[94] Zhou P, Tachedjian M, Wynne JW, Boyd V, Cui J, Smith I, et al. Contraction of the type I IFN locus and unusual constitutive expression of IFN-α in bats. Proceedings of the National Academy of Sciences of the United States of America. 2016;**113**(10):2696-2701. DOI: 10.1073/pnas.1518240113

[95] O'Connor KC. Bats are "blind" to the deadly effects of viruses. Sci Immunol. 2018;**3**(24). DOI: 10.1126/sciimmunol.aau2259

[96] Olival KJ, Hayman DTS. Filoviruses in bats: Current knowledge and future directions. Viruses. 2014;**6**(4):1759-1788. DOI: 10.3390/v6041759

[97] Voigt CC, Kingston T. Bats in the Anthropocene: Conservation of Bats in a Changing World. New York: Springer Nature; 2016. p. 606. DOI: 10.1007/978-3-319-25220-9

[98] Banerjee BD, Seth V, Ahmed RS. Pesticide-induced oxidative stress : Perspective and trends. Reviews on Environmental Health. 2001;**16**(1):1-40. DOI: 10.1515/reveh.2001.16.1.1

[99] Bolognesi C. Genotoxicity of pesticides: A review of human biomonitoring studies. Mutation Research, Reviews in Mutation Research. 2003;**543**(3):251-272. DOI: 10.1016/s1383-5742(03)00015-2

[100] Corsini E, Sokooti M, Galli C, Moretto A, Colosio C. Pesticide induced immunotoxicity in humans: A comprehensive review of the existing evidence. Toxicology. 2013;**307**:123-135. DOI: 10.1016/j.tox.2012.10.009

[101] Repetto R, Baliga SS. Pesticides and immunosuppression: The risks to public health. Health Policy and Planning. 1997;**12**(2):97-106. DOI: 10.1093/heapol/12.2.97

[102] Oliveira JM, Brinati A, Miranda LDL, Morais DB, Zanuncio JC, Gonçalves RV, et al. Exposure to the insecticide endosulfan induces liver morphology alterations and oxidative stress in fruit-eating bats (Artibeus lituratus). International Journal of Experimental Pathology. 2017;**98**(1):17-25. DOI: 10.1111/iep.12223

[103] Oliveira JM, Losano NF, Condessa SS, de Freitas RMP, Cardoso SA, Freitas MB, et al. Exposure to deltamethrin induces oxidative stress and decreases of energy reserve in tissues of the Neotropical fruit-eating bat Artibeus lituratus. Ecotoxicology and Environmental Safety. 2018;**148**:684-692. DOI: 10.1016/j.ecoenv.2017.11.024

[104] Schneeberger K, Czirják GÁ, Voigt CC. Frugivory is associated with low measures of plasma oxidative stress and high antioxidant concentration in free-ranging bats. Naturwissenschaften. 2014;**101**(4):285-290. DOI: 10.1007/s00114-014-1155-5

[105] Agrawal A, Sharma B. Pesticides induced oxidative stress in mammalian systems. International Journal of

Biological and Medical Research. 2010;**1**(3):90-104

[106] Lauridsen C. From oxidative stress to inflammation: Redox balance and immune system. Poultry Science. 2019;**98**(10):4240-4246. DOI: 10.3382/ps/pey407

[107] Moro AM, Brucker N, Charão M, Bulcão R, Freitas F, Baierle M, et al. Evaluation of genotoxicity and oxidative damage in painters exposed to low levels of toluene. Mutation Research, Genetic Toxicology and Environmental Mutagenesis. 2012;**746**(1):42-48. DOI: 10.1016/j.mrgentox.2012.02.007

[108] Liu X, Wu J, Shi W, Shi W, Liu H, Wu X. Lead Induces Genotoxicity via Oxidative Stress and Promoter Methylation of DNA Repair Genes in Human Lymphoblastoid TK6 Cells. Medical Science Monitor. 2018;**24**: 4295-4304. DOI: 10.12659/MSM.908425

[109] Hayashi M. The micronucleus test—most widely used in vivo genotoxicity test—. Genes and Environment. 2016;**38**(1):18. DOI: 10.1186/s41021-016-0044-x

[110] Fairbairn DW, Olive PL, O'Neill KL. The comet assay: A comprehensive review. Mutation Research/Reviews in Genetic Toxicology. 1995;**339**(1):37-59. DOI: 10.1016/0165-1110(94)00013-3

[111] Sandoval-Herrera N, Paz Castillo J, Herrera Montalvo LG, Welch KC Jr. Micronucleus test reveals genotoxic effects in bats associated with agricultural activity. Environmental Toxicology and Chemistry. 2021;**40**(1):202-207. DOI: 10.1002/etc.4907

[112] Benvindo-Souz M, Borges RE, Pacheco SM, de Souza Santos LR. Micronucleus and other nuclear abnormalities in exfoliated cells of buccal mucosa of bats at different trophic levels. Ecotoxicology and Environmental Safety. 2019;**172**:120-127. DOI: 10.1016/j.ecoenv.2019.01.051

[113] Naidoo S, Vosloo D, Schoeman MC. Haematological and genotoxic responses in an urban adapter, the banana bat, foraging at wastewater treatment works. Ecotoxicology and Environmental Safety. 2015;**114**:304-311. DOI: 10.1016/j.ecoenv.2014.04.043

[114] Olival KJ, Hosseini PR, Zambrana-Torrelio C, Ross N, Bogich TL, Daszak P. Host and viral traits predict zoonotic spillover from mammals. Nature. 2017;**546**(7660): 646-650. DOI: 10.1038/nature22975

Section 3

White-Nose Syndrome

Chapter 3

The Physiological Ecology of White-Nose Syndrome (WNS) in North American Bats

Craig L. Frank

Abstract

White-nose Syndrome (WNS) is an emergent mycosis in North America that is caused by a severe cutaneous infection with the fungus *Pseudogymnoascus destructans* (*Pd*) during hibernation. *Pseudogymnoascus destructans* (*Pd*) was first observed in North America at a single site during the winter of 2006–2007 and has since spread to 39 U.S. States and 7 Canadian provinces. This fungus was introduced to North America from Europe, where it is endemic. WNS has thus far been observed to occur only in hibernating bats and has caused the populations of 4 North American bat species to decline by more than 84% within 7 years. Field studies have revealed that 4 other North American bat species are not afflicted with WNS when hibernating in areas where *Pd* occurs. The physiological and biochemical adaptations that permit some bat species to resist *Pd* infections are starting to be elucidated but are still poorly understood. A total of 47 different bat species are found in North America, about half of which hibernate during the winter. The potential future effects of WNS on 13 of these hibernating bat species remains to be determined.

Keywords: mycosis, hibernation, immunity, *Pseudogymnoascus destructans*, lipids

1. Introduction

White-nose Syndrome (WNS) is an emergent mycosis that affects some bat species in North America and is caused by an extensive cutaneous infection with the fungus *Pseudogymnoascus destructans* (*Pd*) during hibernation. It was first observed at a single cave in New York State during the winter of 2006–2007, and then spread to 5 more caves/mines in New York State during the winter of 2007–2008 [1]. *Pseudogymnoascus destructans* (*Pd*) has since spread to 39 U.S. States and 7 Canadian provinces, and it was introduced to North America from Europe [2]. This fungus grows on the muzzle, wings, and ears of afflicted bats during hibernation, with hyphae penetrating both the epidermis and dermis, consuming hair follicles, sebaceous and sweat glands [3–5]. *Pseudogymnoascus destructans* grows at ambient temperatures ranging from 1.9 to 15°C, although the hyphal morphology of this fungus exhibits heat stress at an ambient temperature (T_a) > 12°C, and growth ceases altogether at T_a > 19°C [6]. The arrival of *Pd* has led to severe population declines for 4 of the 6 bat species that hibernate in the Northeastern United States and Canada. Within 1–2 years after the arrival of *Pd* at a hibernation site, the number of little brown (*Myotis lucifugus*), northern long-eared (*Myotis septentrionalis*), Indiana

 IntechOpen

(*Myotis sodalis*), and tricolored (*Perimyotis subflavus*) bats decreases by 75–95% due to high over-winter mortality during the hibernation period [7]. Extensive cutaneous infections with *Pd* have been shown to be the cause of WNS in laboratory inoculation/hibernation experiments with captive *M. lucifugus* [8]. Bats suffering from WNS have numerous skin lesions caused by *Pd* infections on their wings, face, and ears. These *Pd* lesions display an orange-yellow fluorescence when illuminated by long-wave (365–385 nm) UV light [9]. The mechanism by which an extensive cutaneous infection with *Pd* leads to WNS and subsequent death during hibernation in some North American bat species is due to the effects of these infections on the hibernation energetics of small bats.

2. Hibernation physiology and white-nose syndrome

Mammals and birds are unique among animals in that they are homeothermic endotherms, maintaining a constant core body temperature (T_b) over a wide range of ambient temperatures (T_a) through a high metabolic rate [10]. Prolonged periods of high metabolic heat production by mammals and birds requires high rates of energy intake. Food availability in the wild often fluctuates, and consequently the energetic costs of maintaining a high T_b (32–42°C) via endothermy becomes prohibitively expensive during certain environmental conditions. Not all mammals and birds are permanently homeothermic, but instead enter periods of torpor [11]. Torpor is a period when metabolic rate and T_b are greatly reduced. It involves the regulation of T_b at a new and substantially lower level, with a new minimum T_b being maintained. Torpor is a controlled reduction in metabolic rate, T_b, and a suite of physiological processes in endotherms [12]. Mammalian and avian species that employ torpor are therefore classified as heterothermic endotherms [11]. Metabolic rates during torpor can be less than 5% of basal metabolic rate with a corresponding T_b of just 0.5–1.0°C above ambient temperature (T_a) in most instances [13]. Mammals and birds generally employ one of two common patterns of torpor, depending upon species: prolonged torpor during hibernation, and daily torpor. Daily torpor occurs when a heterotherm has torpor bouts that are 3–12 h in duration. Hibernation is seasonal, usually from late summer/autumn to the following spring. Hibernators do not remain torpid continuously throughout the hibernation season; instead, bouts of torpor last from days to weeks, interrupted by brief periods of high metabolic rates and high T_b called arousal episodes. Hibernation is the most common pattern of torpor found in mammals. Numerous studies have revealed that daily torpor occurs in at least 42 bird species, and 78 mammalian species as well. At present, only 1 bird species (the Common Poorwill, *Phalaenoptilus nuttallii*) is known to hibernate, whereas hibernation has been document in about 100 mammalian species [14].

Stored triacylglycerols mobilized from white adipose tissue (WAT) are the primary energy source utilized during mammalian hibernation [15]. The periods of high metabolic rates and T_b known as arousal episodes normally account for 80–90% of stored lipids (energy) utilized during hibernation [16, 17], but their physiological function is poorly understood. A number of physiological/biochemical alterations that occur during torpor are reversed during arousal episodes. These alterations include dendritic retraction, leukocyte sequestration in secondary lymphatic organs, endocytosis, and protein degradation [18]. Periodic arousals from torpor thus serve to rectify physiological/biochemical imbalances that occur during torpor. Heterothermic mammals typically undergo an extensive period of feeding and fattening for several weeks prior to the onset of hibernation, during which body fat levels increase by 4 to 7-fold. The body fat content of *M. lucifugus* increases from 7 to 27% body mass during the 2 months prior to hibernation [19, 20], for example.

The energetic constraints that normal arousal episodes place on the physiological ecology of hibernating bats are illustrated by examining the winter physiological ecology of one bat species that is now severely impacted by WNS, *M. lucifugus*. Although the normal arousal episodes of hibernating *M. lucifugus* are usually less than 1 h in duration, they nonetheless account for 80–90% of all energy utilized during hibernation by this species [21]. Each of these arousal episodes requires the utilization of about 110 mg of stored depot lipids [22]. The body mass of *M. lucifugus* at the onset hibernation averages 8.5 g, indicating that about 2.0 g of depot lipids (triacylglycerols) are stored by each bat to support the entire 190-d hibernation period [23]. If arousal episodes consume a total of at least 80% of the lipids utilized during the entire hibernation period, then about 1.6 g of the 2.0 g stored by *M. lucifugus* are required to fuel them. Hibernating *M. lucifugus* thus have enough stored energy to support only about 14–15 arousal episodes during the entire period. Free-ranging *M. lucifugus* hibernating at ambient temperatures of 5.5 to 12.0°C consequently have normal torpor bouts averaging 12.4 to 19.7 d in length [22, 24] which enables them to survive a 190-d hibernation period without depleting their energy reserves prior to the spring.

Studies on free-ranging *M. lucifugus* revealed that bats with extensive *Pd* infections arouse more frequently from torpor during hibernation, and consequently their torpor bouts were much shorter than the normal range of 12.4 to 19.7 d previously reported for this species. Individuals with extensive cutaneous *Pd* infections (lesions) had a mean (± SE) torpor bout duration of 7.93 ± 2.49 d between arousal episodes, whereas those with no *Pd* lesions had a mean torpor duration of 16.32 ± 6.65 d which was significantly longer [25]. This 51% reduction in torpor bout duration produced by extensive cutaneous *Pd* infections made arousal episodes more frequent, which increased the rate of energy expenditure during the entire hibernation period. This increased rate of energy expenditure during hibernation is WNS, which leads to the premature depletion of body fat reserves prior to the normal spring emergence from hibernation when food (arthropods) first becomes available, which in turn causes of death [26]. The mechanism by which a severe cutaneous infection with *Pd* increases the frequently of arousal episodes during hibernation is related to the degree of evaporative water loss from the skin surface. The normal rate of evaporative water loss (EWL) of bats is considerable during torpor, and it is thought that they periodically arouse from torpor to drink in order to restore their water balance [27]. Hibernating bats have been observed drinking during arousal episodes [28]. It thus has been proposed that the numerous skin lesions caused by severe *Pd* infections may increase the EWL of affected bats, which in turn would cause them to arouse from torpor more frequently to drink. Analyses of blood samples collected from both *Pd* infected and uninfected *M. lucifugus* during hibernation support this hypothesis [29, 30]. Laboratory inoculation/hibernation experiments with *M. lucifugus* revealed that the mean EWL rate of individuals with numerous cutaneous *Pd* lesions was 1.6-fold greater than that for bats with no *Pd* lesions [31]. Interpreting these findings together reveals that when cutaneous *Pd* infections result in numerous skin lesions, WNS is caused by a corresponding increase in the rate of cutaneous EWL, which in turns leads to both reduced torpor bout lengths and more frequent arousal episodes. This subsequently results in a greater rate of energy expenditure during hibernation that produces a premature depletion of body fat reserves during hibernation, before feeding can occur in the spring.

3. The susceptibility to *Pseudogymnoascus destructans*

Field studies on hibernating bats demonstrated that *P. destructans* is found throughout both Europe and Asia, appearing on the skin of these bat species with no apparent increases in over-winter mortality or WNS [32, 33]. The greater mouse-eared bat (*Myotis myotis*) of Europe has been shown to be highly resistant to cutaneous *Pd* infections in both field [34] and laboratory studies [35]. Examination of hibernation sites in Europe revealed *Pd* growing on the muzzles of 5 different European species during torpor: pond (*Myotis dasycneme*), greater mouse-eared bat (*M. myotis*), Daubenton's (*Myotis duabentonii*), Brandt's (*Myotis brandtii*), and lesser mouse-eared (*Myotis oxygnathus*) bats. Mass deaths were not observed at these sites, however, and there were no apparent disruptions in torpor bout duration [34]. Histological analyses of infected *M. myotis* revealed that the hyphae of *Pd* do not extend beyond the epidermis of this bat species, even after several months of hibernation [36]. Some bat species are thus more resistant to *Pd* infections than others, thereby avoiding WNS. Cutaneous *Pd* infections and some associated skin lesions have subsequently been observed in 11 different European and 2 Asia bat species during hibernation [32, 33], with no apparent disruptions in torpor bout length or mortality. These studies indicate that European and Asian species of hibernating bats have evolved a resistance to *Pd* that greatly reduces the extent to which this fungus can infect the skin, thereby reducing the number of lesions that appear during hibernation to the point where torpor bout length is not significantly affected.

A similar resistance to both *Pd* infections and subsequent WNS is displayed by 4 species of North American bats as well. Field studies demonstrated that big brown bats (*Eptesicus fuscus*) hibernating where *Pd* occurs have torpor bouts of normal duration, and usually survive the winter with depot fat remaining [37]. Laboratory hibernation experiments with *E. fuscus* also revealed that *Pd* does not extensively grow in the skin of this species during hibernation [38]. The Eastern small-footed bat (*Myotis leibii*) is also highly resistant to cutaneous infections with *Pd*. A survey of 42 bat hibernation sites in the USA revealed that the number of *M. leibii* at these locations declined on average by only 12% during the first several years since the first appearance of *Pd*, whereas the number of *M. lucifugus*, *M. septentrionalis*, and *P. subflavus* at these sites decreased by 75–98% during this same period [39]. The Southeastern myotis (*Myotis austroriparius*) is a hibernating species found in Southern USA. Examinations of hibernation sites for this species in Alabama reveal although *P. destructans* first appeared in this area during 2011, no increases in the over-winter mortality of *M. austroriparius* have been observed. The skin of 99 hibernating *M. austroriparius* was examined for both the presence *P. destructans* DNA on it, and the UV-florescent skin lesions characteristic of *Pd* infections. Although 77% of the bats tested had *Pd* DNA on their skin, none of them had *Pd* skin lesions [40]. These findings indicate that although *M. austroriparius* was exposed to propagules (spores) of *Pd*, this fungus did not invade the skin of this bat species during hibernation, thus WNS does not develop. The gray bat (*Myotis grisescens*) is listed as an endangered species by the U.S. Fish & Wildlife Service, and some populations hibernate in caves located in the U.S. state of Tennessee. *Pseudogymnoascus destructans* first appeared in these caves during 2013, and some of the *M. grisescens* hibernating in these areas were found to have *Pd* skin lesions. No mass mortality has been observed for this species during hibernation, however, and the number of *M. grisescens* hibernating in these caves has been increasing since 2013. The total number of *M. grisescens* hibernating at 3 Tennessee caves increased by 15.4% during between 2016–2017 and 2018–2019, and it increased by 2.9-fold at another cave during this same period [41]. It therefore appears that the degree of *Pd* infection that occurs during hibernation by *M. grisescens* is not sufficient to cause WNS.

There is ample evidence that some populations of *M. lucifugus* in the Northeastern United States have evolved a greater resistance to cutaneous infections with *Pd*, thereby avoiding WNS. A mark/recapture study conducted by Reichard et al. [42] revealed that the over-winter survival rate of *M. lucifugus* at 8 hibernation sites where *Pd* occurs in the Northeastern USA was at least 5.4%, with some individuals surviving 4 consecutive winters. Another mark and recapture study with *M. lucifugus* hibernating at a site in Michigan revealed that some individuals have survived 7 consecutive winters at a site where *Pd* occurs [43]. Studies on hibernating free-ranging *M. lucifugus* conducted during the first several years after the appearance of *Pd* revealed that although most individuals developed severe *Pd* infections with a high density of skin lesions that resulted in WNS, some individuals developed only moderate *Pd* infections that produced far fewer skin lesions, avoiding WNS and surviving the winter. Furthermore, the mean (± SE) torpor bout duration of these individuals with fewer lesions was 13.96 ± 4.30 d, which was not significantly different from that observed for *M. lucifugus* hibernating with no cutaneous *Pd* infections [25]. The differences in torpor bout lengths between individual *M. lucifugus* hibernating at the same site where *Pd* is found is illustrated by the skin temperature (T_{skin}) recordings of 2 adult females hibernating at the Williams Preserve Mine in New York State during the November–December period of 2008 (**Figure 1**). Skin temperature is equivalent to body temperature in small bats [44]. The first bat (**Figure 1A**) began hibernation with torpor bouts that were normal in length (15–20 d), but torpor bout length decreased to just 7–9 d during December 2008, indicating that this individual had succumbed to WNS. The second *M. lucifugus* (**Figure 1B**) examined, however, maintained torpor bouts that were 15–20 d long throughout the study period, demonstrating that it was not afflicted with WNS.

These studies indicate that for *M. lucifugus*, some individuals within certain populations are more resistant to *Pd* infections than others, and these are the bats that are surviving consecutive winters despite the presence of *Pd*. The consistent survival of some *M. lucifugus* in the presence of *Pd* has led to a partial recovery of some populations in New York State. A small maternity colony of *M. lucifugus* in NY examined by Dobony and Johnson [45] during the summers of 2006 through 2017 demonstrated that the size of it decreased by 88% after the first appearance of *Pd*, then stabilized during 2010–2014, and has been increasing since 2014. The New York Department of Environmental Conservation has been conducting annual counts of hibernating bats at the Williams Preserve Mine and Hailes Cave since 1999. These are 2 of the 6 bat hibernation sites where *Pd* first appeared during the winter 2007–2008. The number of *M. lucifugus* hibernating at the Williams Preserve Mine during the winter of 2008–2009 was just 12% of that observed prior to the first appearance of *Pd*, and the number at Hailes Cave was just 9% of the pre-*Pd* level for this site. The number of *M. lucifugus* observed at these sites during subsequent hibernation periods has since consistently increased, however. The number of *M. lucifugus* increased to 41% of pre-*Pd* levels by 2017 at the Williams Preserve Mine, and increased to 31% of pre-*Pd* levels at Hailes Cave by 2017 [23]. Another field study conducted at the Williams Preserve Mine indicates that this *M. lucifugus* population has evolved a higher resistance to *Pd* growth on their wings [46].

A field study conducted at a single hibernation site in NY during the winter of 2014–2015, about 6 years after *Pd* had arrived, indicated that the mean (± SE) torpor bout duration of *M. lucifugus* surviving the winter was 12.0 ± 10.8 d [47], which

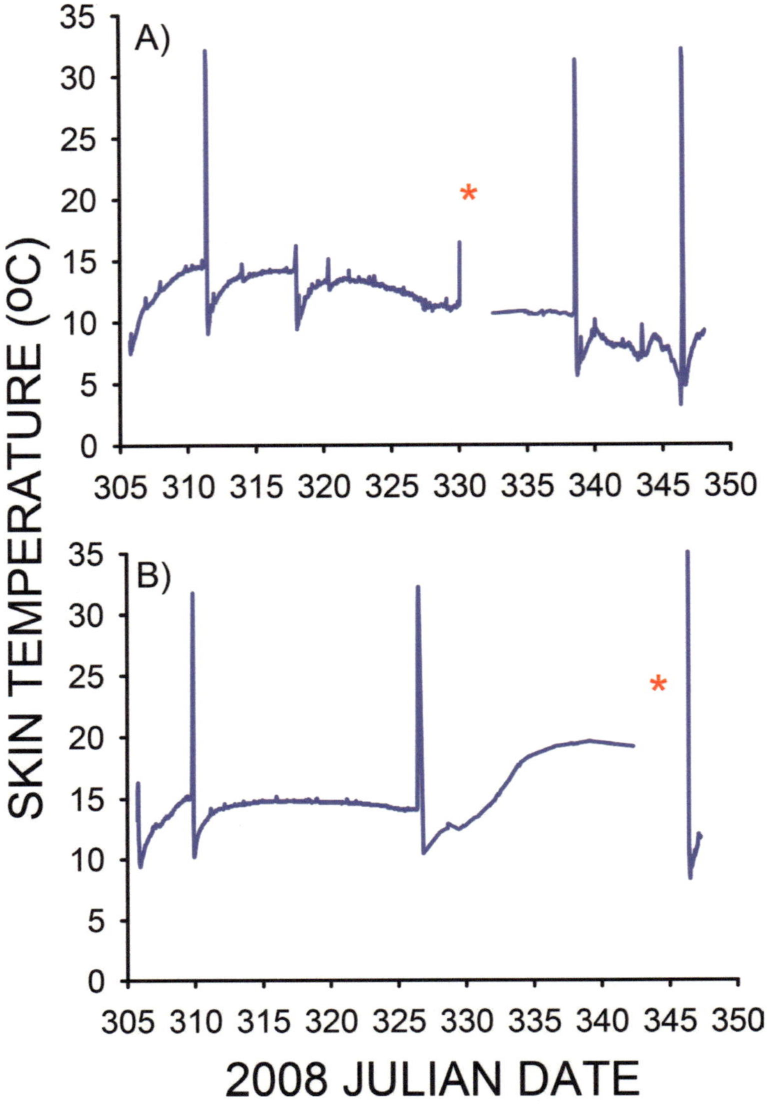

Figure 1.
*(A) Skin temperatures of two different female Myotis lucifugus hibernating at the Williams preserve mine during November and December 2008. (B) Recordings of T_{skin} began 45 min after a temperature-sensitive radio transmitter was placed on each bat during Julian day 305. Radio signals were continuously recorded at 10–15 min intervals, torpor is defined as when T_{skin} < 24°C. *indicates a period when the bat was out of the range of the automated radio receiver during an arousal episode. Data are from Frank et al. [23].*

is close to the normal torpor bout duration of 15–20 d previously reported for this species, thus indicating that most were hibernating normally. Another field study on *M. lucifugus* hibernating in the Williams Preserve Mine revealed that the bats hibernating at this site 1 year after the arrival of *Pd* (2008–2009) had: a) a mean torpor bout duration of 7.6 d, b) no depot fat reserves remaining by March, and c) an apparent over-winter mortality rate of 88%. The *M. lucifugus* hibernating at this same site 6–9 years after the arrival of *Pd*, in contrast, had: a) a mean torpor bout duration of 14.7 d, b) depot fat remaining in March, and c) an apparent over-winter mortality rate of 50% [23]. Interpreting these studies together reveals that some populations of *M. lucifugus* have recently evolved a greater resistance to cutaneous infections with *Pd*, thus reducing the frequency of WNS.

4. Impacts of WNS

The impact of WNS on the total number of *M. lucifugus* in North America has been drastic, despite the partial recovery of some populations in New York State. A recent analysis of winter hibernaculum counts for 5 bat species performed in 27 U.S. states and 2 Canadian over a 23-year period was performed by Cheng et al. [48]. Their analysis revealed that the total number of *M. lucifugus*, *M. septentrionalis*, and *P. subflavus* hibernating at these sites decreased by over 90% with the first 7 years after the arrival of *Pd*, and the number of *M. sodalis* decreased by 84%. The decline in the number of *M. septentrionalis* was so great that the U.S. Fish and Wildlife Service designated it a threatened species in 2016. A total of 47 different bat species are found in North America, and about half of them are known to hibernate during the winter [49]. Although the susceptibility of 8 of these hibernating bat species to WNS is well understood, the ability of *Pd* infections to cause WNS in other species of hibernating North American bats is virtually unknown. The National Wildlife Health Center of the U.S. Geological Survey thus began a WNS surveillance program in 2013 to address the potential effects of *Pd* as it spreads across the USA. Each year the wildlife agencies of 22 U.S. states submit thousands skin surface swabs collected from free-ranging bats during either the fall or hibernation to the National

Scientific name	Common name	*Pd* susceptibility
Myotis lucifugus	Little brown myotis	WNS
Myotis septentrionalis	N. Long-eared myotis	WNS
Myotis sodalis	Indiana myotis	WNS
Perimyotis subflavus	Tricolored bat	WNS
Myotis californicus	California myotis	Unknown
Myotis thysanodes	Fringed myotis	Unknown
Myotis evotis	W. Long-eared myotis	Unknown
Antrozous pallidus	Pallid bat	Unknown
Parastrellus hesperus	Canyon bat	Unknown
Myotis auriculus	Southwestern myotis	Unknown
Idionycteris phyllotis	Allen's big-eared bat	Unknown
Myotis leibii	Small-footed myotis	No WNS
Myotis austroriparius	Southeastern myotis	No WNS
Eptesicus fuscus	Big brown bat	No WNS
Myotis grisescens	Gray myotis	No WNS
Myotis ciliolabrum	Small-footed myotis	*Pd* DNA
Corynorhinus townsendii	Townsend's bat	*Pd* DNA
Lasionycteris noctivagans	Silver-haired bat	*Pd* DNA
Myotis yumanensis	Yuma myotis	*Pd* infection
Myotis volans	Long-legged myotis	*Pd* infection
Myotis velifer	Cave myotis	*Pd* infection

Table 1.
North American species of hibernating bats and their known levels of susceptibility to cutaneous infection with Pseudogymnoascus destructans.

Wildlife Health Center which are examined for *Pd* DNA. Skin biopsies collected from bats showing signs of fungal infection are also submitted for histological analyses [A. Ballman, personal communication]. The result of these surveys are summarized in the WNS webpage maintained by the U.S. Fish & Wildlife Service [50]. The degree to which each species of hibernating North America bat is susceptible to cutaneous *Pd* infections and WNS has been compiled in **Table 1** using both the results of both the surveys conducted by National Wildlife Health Center as well as the published studies cited previously. The susceptibility of 7 of the 21 bat species listed in **Table 1** to *Pd* infections during hibernation is currently unknown due to a lack of data for these species. The DNA of *Pd* has been found on the skin surface of *Myotis ciliolabrum*, *Corynorhinus townsendii*, and *Lasionycteris noctivagans* indicating that *Pd* now occurs with the range of these species, but it is not known if *Pd* can invade their skin and produce lesions. Skin lesions caused by *Pd* infections have been documented in hibernating *Myotis yumanensis*, *M. volans*, and *Myotis velifer*, but it is not known if the density of *Pd* lesions that appear during this period is sufficient to cause WNS. The degree of *Pd* infection that occurs in *Myotis austroriparius*, *M. grisescens*, *Myotis leibii*, and *E. fuscus* is not sufficient to cause WNS in these species.

5. Cutaneous lipids

Four of the hibernating bat species found in North America as well as 11 hibernating bat species in Europe limit the degree of cutaneous *Pd* infection to the point where it does not result in WNS during hibernation. The physiological and biochemical mechanisms that enable these bat species to reduce *Pd* infections are poorly understood. Several recent studies have revealed that one factor which confers a resistance to *Pd* infections is the lipid composition of the epidermis. The outermost stratum of the epidermis is the first defense against fungal skin infections because the mycelium must initially invade it, and leukocytes are not present in these epidermal layers. The epidermis is composed chiefly of special epithelial cells called keratinocytes that occur in 4 strata; they are produced in the deepest stratum (the stratum basale) and migrate to the top stratum (the stratum corneum) as they age. Epidermal surface lipids are a mixture of compounds secreted by keratinocytes into the intracellular matrix, and sebum secreted onto it by the sebaceous glands. The lipid mixture secreted by keratinocytes contains free sphingosine bases, ceramides, cholesterol, and free fatty acids (FFAs), whereas the sebum is composed of triacylglycerols, diacylglycerols, FFAs, wax esters, squalene, cholesterol, and cholesterol esters [51, 52]. The epidermal lipids of bats also have cerebrosides and monoacylglycerols [53, 54]. Some free fatty acids (FFAs) are known to have antimicrobial effects [55].

It has been demonstrated that the wing epidermis of both *M. lucifugus* and *E. fuscus* contain the same 7 fatty acids: Myristic (14:0), pentadecanoic (15:0), palmitic (16:0), palmitoleic (16:1), stearic (18:0), oleic (18:1), and linoleic (18:2) acids. The wing epidermis of hibernating *E. fuscus* contains about twice as much myristic, palmitoleic, oleic, and linoleic acids than that of *M. lucifugus*, as well. Laboratory experiments with *Pd* cultures revealed that pentadecanoic, palmitoleic, oleic, and linoleic acids in the free fatty acid (FFA) form inhibit the growth of *Pd* [56, 57], with linoleic acid reducing it by more than 99%. The results of one of these experiments are summarized in **Figure 2**.

Epidermal free fatty acid composition thus appears to be one of the factors that enables *E. fuscus* to better resist *Pd* infections than *Myotis lucifugus*.

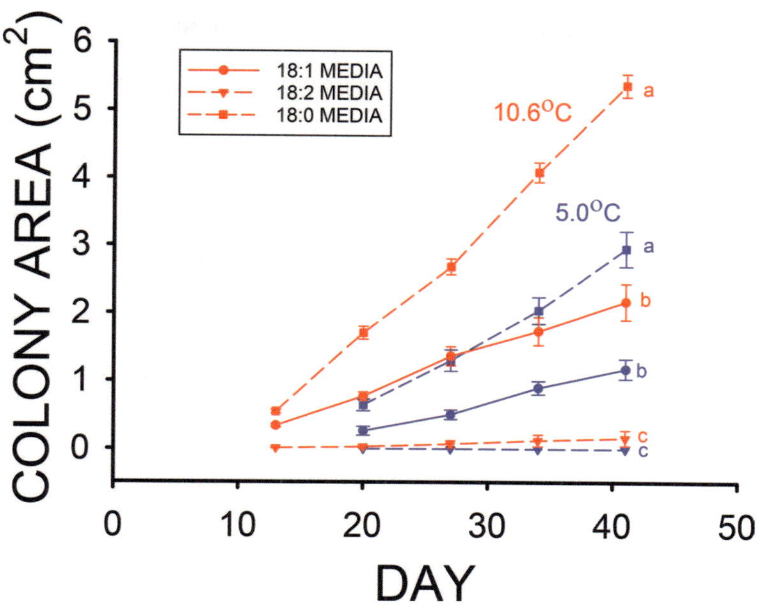

Figure 2.
Mean (± SE) surface areas of Pd colonies at various growth stages on media containing 1% of either oleic (18:1), linoleic (18:2) or stearic (18:0) acids, while being incubated at 5.0 (blue symbols) and 10.6°C (red symbols). Mean within the same temperature treatment sharing a common lower-case letter are not significantly different at the P < 0.05 level. Data are from Frank et al. [56].

Wax esters consist of an alcohol linked to a fatty acid molecule with an ester bond [58]. About 120 different wax esters have been found in the sebum of *Myotis myotis* during hibernation [59]. A recent study used laboratory *Pd* culture experiments to determine the effects of some of these wax esters on *Pd* growth [60]. These experiments have revealed that 4 of the wax esters found in the sebum of *M. myotis* inhibit *Pd* growth by over 90%. These anti-*Pd* wax esters are: behenyl linoleate, palmityl linoleate, arachidyl linoleate, and behenyl palmitoleate. One factor that enables *M. myotis* to resist *Pd* infections is therefore presence of these anti-*Pd* wax esters in their epidermis. Changes in epidermal lipid composition may also be one of the adaptations that permit some populations of *M. lucifugus* to now have a higher resistance to cutaneous *Pd* infections.

6. Conclusions

Four species of North American bats develop severe *Pd* infections during hibernation which result in WNS, whereas 4 other bat species develop only moderate *Pd* infections during this period and do not display the abnormal torpor patterns and mortality associated with WNS. It has been also demonstrated that within the same population of *M. lucifugus*, individuals with a high density of *Pd* lesion in their skin suffer from WNS, whereas those with a much lower density of *Pd* lesions do not. Cutaneous infections with *Pd* thus do not always result in WNS, and a certain density of *Pd* lesions must be present in the skin before WNS develops. The precise minimum *Pd* lesion density for WNS is not known, and it should be determined to better understand both the ecology and potential impacts of WNS on the hibernating bat populations of North America, as the mere presence of some *Pd* lesions alone does not indicate that a bat species/population is afflicted with WNS. The effects of these lesions on torpor bout length must be taken into consideration when assessing potential WNS. The full extent to which *Pd* can infect the skin and causes WNS in 13 of 21 bat species listed in **Table 1** is not known. Most of these 13 bat species occur in Western North America. It is thus unclear how WNS will affect most Western bat species as *Pd* spreads across North America. If current trends in *Pd* susceptibility continue, then it is likely that WNS will severely affect 7–8 additional bat species in North America. Analyses of preserved museum specimens suggest that *Pd* has been associated with hibernating bats in Europe for over 100 years, whereas they reveal no evidence that *Pd* was present in North America bats between 1861 and 1971 [61]. It therefore appears that bats across Europe have adapted to *Pd* for over 100 years, and they are now highly resistant to *Pd* infections, thereby avoiding WNS. The physiological/biochemical basis of this resistance is largely unknown but warrants further investigation to better predict which New World bat species will be severely affected by WNS.

Acknowledgements

I thank April Davis, Carl Herzog, Alan Hicks, and Melissa Ingala for helpful discussions which greatly improved this review.

Conflict of interest

The author declares no conflict of interest.

Author details

Craig L. Frank
Department of Biological Sciences, Fordham University, Armonk, NY, USA

*Address all correspondence to: frank@fordham.edu

References

[1] Keller S, Lorch JM, Berlowski-Zier BM, Ballman A, Blehert DS. Analysis of archival specimens confirms White-Nose Syndrome in little brown bats (*Myotis lucifugus*) from New York, USA, in spring 2007. Journal of Wildlife Diseases. 2021;57 (2): 457-460. DOI:DOI:10.7589/JWD-D-20-00137

[2] Trivedi J, Lachapelle J, Vanderwolf KJ, Misra V, Willis CKR, Ratcliffe JM, et al. Fungus causing white-nose syndrome in bats accumulates genetic variability in North America with no sign of recombination. mSphere. 2017;2:e00271-e00217. DOI:10.1128/mSphereDirect.00271-17.

[3] Blehert DS, Hicks AC, Behr M, Meteyer CU, Berlowski-Zier BM, Buckles EL, et al. Bat white-nose syndrome: an emerging fungal pathogen? Science. 2009;323(5911):227.

[4] Gargas A, Trest MT, Christensen M, Volk TJ, Blehert DS. *Geomyces destructans* sp. Nov. associated with bat white nose syndrome. Mycotaxon. 2009;108:147-154.

[5] Chaturvedi V, Springer DJ, Behr MJ, Ramani R, Li, X, Peck MK, et al. Morphological and molecular characterizations of psychrophilic fungus *Geomyces destructans* from New York bats with White Nose Syndrome (WNS). PLoS ONE. 2010;5(5):e10783.

[6] Verant ML, Boyles JG, Waldrep W Jr, Wibbelt G, Blehert DS. Temperature-dependent growth of *Geomyces destructans*, the fungus that causes bat white-nose syndrome. PLoS ONE. 2012;7(9):e46280. DOI:10.1371/journal.pone.0046280

[7] Langwig KE, Hoyt JR, Parise KL, Kath J, Kirk D, Frick WF, et al. Invasion dynamics of white-nose syndrome fungus, midwestern United States. Emerging Infectious Diseases. 2015;21:1023-1026.

[8] Warnecke L, Turner JM, Bollinger TK, Lorch JM, Misra V, Cryan PM, Wibbelt G, Blehert DS, Willis CK. Inoculation of bats with European *Geomyces destructans* supports the novel pathogen hypothesis for the origin of White-nose syndrome. PNAS. 2012;109:6999-7003.

[9] Turner GG, Meteyer CU, Barton H, Gumbs JF, Reeder DM, Overton B, et al. Nonlethal screening of bat-wing skin with the use of ultraviolet fluorescence to detect lesions indicative of white-nose syndrome. Journal of Wildlife Diseases. 2014;50(3):566-573.

[10] Hill RW, Wyse GA, Anderson M. Animal Physiology, 4th ed. New York: Sinauer Associates; 2016. 4-0 p.

[11] Geiser, F. Metabolic rate and body temperature. Annual Review of Physiology. 2004;66:239-274.

[12] Somero GN, Lockwood BL, Tomanek L. Biochemical Adaptation: Response to Environmental Challenges from Life's Origins to the Anthropocene. New York: Oxford University Press; 2017. 572 p.

[13] Geiser F, Ruf T. Hibernation versus Daily Torpor in Mammals and Birds: Physiological variables and classification of torpor patterns. Physiological Zoology. 1995 68(6), 935-966.

[14] Ruf T, Geiser F. Daily torpor and hibernation in birds and mammals. Biological Reviews. 2015;90 891-926. DOI:10.1111/brv.12137

[15] Dark J. Annual lipid cycles in hibernators: integration of physiology and behavior. Annual Review of Nutrition. 2005;25: 469-497.

[16] Kayser C. Hibernation. In: Mayer W, Van Gelder R, editors. Physiological Mammalogy, volume II. New York: Academic Press; 1965. p. 180-296.

[17] Karpovich SA, Tøien Ø, Buck CL, Barnes BM. Energetics of arousal episodes in hibernating arctic ground squirrels. Journal of Comparative Physiology B. 2009;179:691-700.

[18] van Breukelen, F, Martin SL. The Hibernation Continuum: physiological and molecular aspects of metabolic plasticity in Mammals. Physiology. 2015;30: 273-281. DOI:10.1152/physiol.00010.2015

[19] Kunz TH, Wrazen JA, Burnett CD. Changes in body mass and body composition in pre-hibernating little brown bats (*Myotis lucifugus*). Ecoscience. 1998;5:8-17.

[20] Reynolds RS, Kunz TH. Changes in body composition during reproduction and postnatal growth in the little brown bat, *Myotis lucifugus* (Chiroptera: Vespertilionidae). Ecoscience. 2000;7: 10-17.

[21] Thomas DW. The physiological ecology of hibernation in vespertilionid bats. Symposia of the Zoological Society of London. 1995;67:233-244.

[22] Thomas DW, Dorais J, Bergeron J-M. Winter energy budgets and cost of arousals for hibernating little brown bats, *Myotis lucifugus*. Journal of Mammalogy. 1990;71:475-479.

[23] Frank CL, Davis AD, Herzog, C. The evolution of a bat population with white-nose syndrome (WNS) reveals a shift from an epizootic to an enzootic phase. Frontiers in Zoology. 2019;16:40. DOI:10.1186/s12983-019-0340-y

[24] Brack V, Twente JW. The duration of the period of hibernation of three species of vestertilionid bats I. Field studies. Canadian Journal of Zoology. 1985;63:2952-2954.

[25] Reeder DM, Frank CL, Turner GC, Meteyer CU, Kurta A, Britzke ER, et al. Frequent arousal from hibernation linked to severity of infection and mortality in bats with White-nose Syndrome. PLoS ONE. 2012;7(6):e38920. DOI:10.1371/journal.pone.0038920

[26] Frank CL, Diaz P, Kunz TH. The relationship between White-nose Syndrome and dietary PUFA levels in bats. In: Ruf T, Bieber C, Arnold W, Millesi E, editors. Living in a seasonal world: thermoregulatory and metabolic adaptations. Germany: Springer-Verlag; 2012. p. 271-279.

[27] Thomas DW, Geiser F. Periodic arousals in hibernating mammals: is evaporative water loss involved? Functional Ecology. 1997;11:585-591.

[28] Wilcox A, Warnecke L, Turner JM, McGuire LP, Jameson JW, Misra V, Bollinger TC, Willis CK. Behaviour of hibernating little brown bats experimentally inoculated with the pathogen that causes White-nose Syndrome. Animal Behavior. 2014;88:157-164.

[29] Cryan PM, Meteyer CU, Blehert DS, Lorch JM, Reeder DM, Turner GG, Webb J, Behr M, Verant M, Russell RE, Castle KT. Electrolyte depletion in white-nose syndrome bats. Journal of Wildlife Diseases. 2013;49(2):398-343.

[30] Warnecke L, J. Turner JM, Bollinger TK, Misra V, Cryan PM, Blehert DS, Willis CK. Pathophysiology of white-nose syndrome in bats: a mechanistic model linking wing damage to mortality. Biology Letters. 2013;9:20130177.

[31] McGuire LP, Mayberry HW, Willis CKR. White-nose syndrome increases torpid metabolic rate and

evaporative water loss in hibernating bats. American Journal of Physiology. 2017;313:R680-R686. DOI:10.1152/ajpregu.00058.2017

[32] Zukal, J. et al. White-nose syndrome without borders: *Pseudogymnoascus destructans* infection tolerated in Europe and Palearctic Asia but not in North America. Scientific Reports. 2016;6:19829; DOI:10.1038/srep19829 (2016).

[33] Hoyt JR, Sun K, Parise KL, Lu G, Langwig KE, Jiang T, et al. Widespread bat white-nose syndrome fungus, Northeastern China. Emerging Infectious Disease. 2016;22(1):140-142. DOI:10.3201/eid2201.151314

[34] Wibbelt G, Kurth D, Hellmann M, Weishaur A, Barlow M, Veith J, Pruger T, Gorfol L, Grosche F, Bontadina V, Zophel HP, Seidl P, Cryan PM, Belhert DS. White-nose syndrome fungus (*Geomyces destructans*) in bats, Europe. Emerging Infectious Diseases. 2010;16:1237-1242.

[35] Davy C, Donaldson M, Willis CKR, Saville B, McGuire L, Mayberry H, Wilcox A, Wibbelt G, Misra V, Bollinger T, Christopher JK. The other white-nose syndrome transcriptome: Tolerant and susceptible hosts respond differently to the pathogen Pseudogymnoascus destructans. Ecology and Evolution. 2017;7: 10.1002/ece3.3234.

[36] Wibbelt G, Puechmaille SJ, Ohlendorf B, Mu¨ hldorfer K, Bosch T, et al. Skin lesions in European hibernating bats associated with *Geomyces destructans*, the etiologic agent of White-Nose Syndrome. PLoS ONE 2013;8(9): e74105. DOI:10.1371/journal.pone.0074105

[37] Frank CL, Michalski A, McDonough AA, Rahimian M, Rudd RJ, Herzog CJ, et al. The resistance of a North American bat species (*Eptesicus fuscus*) to White-nose Syndrome (WNS). PLoS ONE. 2014;9(12):e113958; DOI:10.1371/journal.pone.0113958.

[38] Moore MS, Field KA, Behr MJ, et al. Energy conserving thermoregulatory patterns and lower disease severity in a bat resistant to the impacts of white-nose syndrome. Journal of Comparative Physiology B. 2018;88: 163. DOI:10.1007/s00360-017-1109-2.

[39] Turner GG, Reeder D, Coleman J. A five-year assessment of mortality and geographic spread of white-nose syndrome in North American bats and a look to the future. Bat Research News. 2011;52:13-27.

[40] Johnson JS, Sharp NW, Monarchino MN, Lilley TM, Edelman AJ. No Sign of infection in free-ranging *Myotis austroriparius* hibernating in the presence of *Pseudogymnoascus destructans* in Alabama. Southeastern Naturalist. 2021;20(1):20-28. DOI:10.1656/058.020.0102

[41] Campbell J. Tennessee winter bat population and White-nose Syndrome monitoring report for 2018-2019. TWRA Technical Report. 2019;19:6

[42] Reichard JD, Fuller NW, Bennett AB, Darling SR, Moore MS, Langwig KE, et al. Interannual survival of *Myotis lucifugus* (Chiroptera: Vespertilionidae) near the epicenter of White-nose Syndrome. Northeastern Naturalist. 2014;21(4):N56-N59.

[43] Kurta A, Foster RW, Daly BA, Wilson AK, Slider RM, Rockey CD, Rockey JM, Long BL, Auteri GG, Collins JD, White JP, Kaarakka HM, Redell JA, Reeder DM. Exceptional longevity in little brown bats still occurs, despite presence of White-Nose Syndrome. Journal of Fish and Wildlife Management. 1 2020;11(2): 583-587. DOI:DOI:10.3996/JFWM-20-039

[44] Willis CKR, Brigham RM. Defining torpor in free-ranging bats: experimental evaluation of external temperature sensitive radio transmitters and the concept of active temperature. Journal of Comparative Physiology B. 2003;173:379-389.

[45] Dobony CA, Johnson JB. Observed resiliency of little brown myotis to long-term white-nose syndrome exposure. Journal of Fish and Wildlife Management. 2018;9(1):168-179.

[46] Langwig KE, Hoyt JR, Parise KL, Frick WF, Foster JT, Kilpatrick AM. Resistance in persisting bat populations after white-nose syndrome invasion. Philosophical Transactions of the Royal Society B. 2017;372:20160044.

[47] Lilley T, Johnson J, Ruokolainen L, Rogers E, Wilson C, Mead Schell S, et al. White-nose syndrome survivors do not exhibit frequent arousals associated with *Pseudogymnoascus destructans* infection. Frontiers in Zoology. 2016;13. 10.1186/s12983-016-0143-3.

[48] Cheng TL, et al. The scope and severity of white-nose syndrome on hibernating bats in North America Conservation Biology. 2021;1-12. DOI:10.1002/cobi.13739.

[49] Feldhamer GA, Drickamer LC, Vessey SH, Merrit JF, Krajewski C. Mammalogy, Adaptation, Diversity, Ecology. New York: McGraw Hill; 2007. 625 p.

[50] Bats Affected by WNS [Internet]. 2021. Available from: https://www.whitenosesyndrome.org/static-page/bats-affected-by-wns [Accessed: 2021-09-01]

[51] Camera E; Ludovici M; Galante M; Singagra J-L; Picardo M. Comprehensive analysis of the major lipid classes in sebum by rapid resolution high performance liquid chromatography and electrospray mass spectrometry. Journal of Lipid Research. 2010;51: 3377-3388.

[52] Feingold KK. The role of epidermal lipids in cutaneous permeability barrier homeostasis. Journal of Lipid Research. 2007;48: 2531-2546.

[53] Ben-Hamo M, Munoz-Garcia A; Larrain, P; Pinshaw B, Korine C, Williams JB. The cutaneous lipid composition of bat wing and tail membranes: a case of convergent evolution with birds. Proceedings of the Royal Society B. 2016;283:20160636. DOI:10.1098/rspb.2016.0636.

[54] Pannkuk EL, Gilmore DE, Fuller NW, Savary BJ, Risch TS. Sebaceous lipid profiling of bat integumentary tissues: quantitative analysis of free fatty acids, monacylglycerides, squalene and sterols. Chemistry and Biodiversity. 2013;10:2122-2132.

[55] Fisher CL, Blanchette DR, Brogden KA, Dawson DV, Drake DR, Hill JR, Wertz PW. The roles of cutaneous lipids in host defense. Biochimica et Biophysica Acta. 2014;1841(3): 319-322.

[56] Frank CL, Ingala MR, Ravenelle RE, Dougherty-Howard K, Wicks SO, Herzog C, et al. The effects of cutaneous fatty acids on the growth of *Pseudogymnoascus destructans*, the etiological agent of White-nose Syndrome (WNS). PLoS ONE. 2016;11(4):e0153535. DOI:10.1371/journal.pone.0153535.

[57] Ingala, MR, Monro, JJ, Ravenelle, RE, Frank CL. The effects of epidermal fatty acid profiles, 1-oleoglycerol, and triacylglycerols on the susceptibility of hibernating bats to Pseudogymnoascus destructans. PLoS ONE. 2017;12(10):e0187195. DOI:10.1371/journal.pone.0187195.

[58] Gunstone FD. Fatty Acid and Lipid Chemistry. New York: Chapman and Hall; 1996. 375 p.

[59] Řezanka T, Viden I, Nováková A, Bandouchová H, Sigler K. Wax ester analysis of bats suffering from White Nose Syndrome in Europe. Lipids. 2015;50:633-645. DOI:10.1007/s11745-015-4027-7.

[60] Frank CL, Sitler-Elbel KG, Hudson AJ, Ingala MR. The antifungal properties of epidermal fatty acid esters: insights from White-Nose Syndrome (WNS) in bats. Molecules. 2018;23:1986. DOI:10.3390/molecules23081986

[61] Campana MG, Kurata NP, Foster JT, Helgen LE, Reeder DM, Fleischer RC, et al. White-Nose Syndrome fungus in a 1918 bat specimen from France. Emerging Infectious Diseases 2017;23(9):1611-1612. DOI:10.3201/eid2309.170875

Section 4

Ecosystem Management

Chapter 4

Bats and Ecosystem Management

Kareem M. Soliman and Wiame W. Emam

Abstract

Bats are among the most misperceived and undervalued animals on the planet. For wildlife ecologists, they are wonderful and incredibly fascinating creatures, but people's feelings about bats are often negative, perhaps because bats are so mysterious. Unfortunately, these fears and myths about bats threaten conservation, biodiversity, and the entire ecosystem. Bats are among the most diverse and geographically dispersed group of living mammals. They contribute to several ecosystem services and act as biological pest crop control agents. Their abundance may reflect changes in populations of arthropod prey species. Also, bats have significant potentials as bioindicators that demonstrate measurable responses to climate change and habitat loss and that induce large-scale impacts on the biota. Indeed, bat conservation is fundamental not only for biodiversity, but also because these flying mammals provide essential ecological and economic services to humans.

Keywords: agricultural pests, ecosystem services, bioindicators, environmental stressors

1. Introduction

Many ecosystems worldwide have become more threatened than ever before through many anthropogenic activities such as urbanization, deforestation and different kinds of pollution. Healthy ecosystems are crucial for human wellbeing improvement, since they provide various ecological services (e.g., insect suppression, pollination, seed dispersal, water and air purification, stabilization of soils, decomposition of wastes, binding of toxic substances, mitigation of diseases, mitigation of floods, and regulation of climate, etc.), many provisions (e.g., food, fuel, fiber, and medicines), and cultural benefits (e.g., esthetic, spiritual, educational, and recreational) [1]. These processes and products are commonly referred to as ecosystem services [2, 3] and have been duly recognized by the United Nations Millennium Ecosystem Assessment [4, 5].

Furthermore, ecosystem services mainly depend on the type of ecosystem as well as the organisms that they constitute. In this chapter, the ecosystem services provided by bats are considered. Bats have been on Earth for over 52 million years [6], with over 1400 species worldwide. They are classified as the second-largest mammalian order with great physiological and ecological diversity [7]. Due to their diverse behavioral, roosting, and feeding habits, many species of bats roost during the daytime in foliage, caves, rock crevices, hollows of trees, beneath exfoliating bark, and different manmade structures [8–11]. While during the night, bats fill the sky to forage on a variety of food items ranging from insects, nectar, and fruit, to seeds, frogs, fish, small mammals, and even blood.

2. Bat: human interaction

Bat-human interrelationships are probably as old as humanity itself. However, the oldest bat fossils are estimated to date back to the Eocene, exceeding 50 million years ago [6]. These nocturnal, fast-flying, and secretive mammals may have been as enigmatic to human ancestors as they are to most of us today. The ancient coexistence of people and bats has been translated into enormous historical and contemporary cultural representations of bats in local folklore [12–14]. This ancient connection exquisitely celebrated in abundant reddish-terracotta rock paintings of bats was made by hunter-gatherer colonizers of the northwest Amazon during the Late Pleistocene/Early Holocene transition [15]. Human still considers bats to be loathsome and fearful creatures, despite the countless research investigating the contributions of bats to the ecosystem and the benefits they provide to human wellbeing [16–21].

2.1 Bat's negative stigma

Contrary to popular belief, bats do not attack people; they do not get tangled in people's hair; and even vampire bats are not real vampires (vampire bats lick blood but do not suck blood). Underlying these negative sentiments are relatively pervasive associations between bats and death [22, 23], witchcraft [24–26], vampires [27–29], malevolent spirits [30–32], and evil [33, 34].

Across much of Europe, disdain and fear of bats are largely embroiled in Church doctrine, where bats are often used to symbolize Satan [35]. However, elsewhere (particularly throughout Southeast Asia, China, and Japan), bats are also associated with luck and good fortune [13, 36] and used as spiritual totems [37, 38]. Among several indigenous groups in Mexico, bats have been considered messengers from the underworld and important symbols of fertility [39].

Furthermore, as human numbers increase and people encroach deeper into the remaining natural habitats, human-bat interactions are becoming more frequent, often with undesirable consequences for both humans and bats.

Human-bat conflicts often arise from damage to buildings or as a result of noise/smelly caused by synanthropic species [14, 40] or due to fruit crop invasion by frugivorous species [21, 41]. These are other key areas in which ethnobiological work can substantially contribute to support evidence-based and culturally sensitive strategies aimed at reducing negative feelings toward bats.

2.2 Ecosystem services

An ecosystem is a system consisting of biotic and abiotic components that function together as a unit. The biotic components include all the living things whereas the abiotic components are the non-living things. Thus, an ecosystem science definition entails an ecological community consisting of different populations of organisms that live together in a habitat. Ecology, which is the scientific study of the interactions between populations or between organisms and the environment, can be viewed at the level of an individual, population, community, or ecosystem. Ecology at the level of individuals is mainly concerned with the physiology, reproduction, and development of an organism. At the level of population, ecology deals primarily with the attributes and the various factors affecting the population. At the level of community, ecology investigates the interactions between populations and community patterns. At the ecosystem level, ecology brings them all together to understand how the system functions as a unit [42].

Ecosystem services are the benefits obtained from the environment that increase human well-being. Economic valuation is conducted by measuring the human welfare gains or losses resulting from changes in the provision of ecosystem services. Bats, which live on all continents except Antarctica, are essential members of many types of ecosystems, ranging from rainforests to deserts. By fulfilling their ecosystem roles, bats promote biodiversity and support the health of their ecosystems. Due to the rich diversity of dietary habits of bats, ranging from species that feed on insects and other arthropods to those that feed on fruit, nectar, and flowers, they have long been postulated to play important roles in ecological and economic services. Moreover, other species that feed on seeds, frogs, fish, small mammals, and even blood also play important roles in ecosystems as predators or prey in ecosystems sustainability. Modifying ecosystems to facilitate socioeconomic development is necessary but how can we avoid damaging important ecosystem services? As a prerequisite, we need to understand how ecosystem services contribute to people's livelihoods and well-being.

2.2.1 Ecological services

Bats have long been known as the main contributor to remarkable ecosystem benefits, are significant suppressors of agricultural pests [43–46], consume important disease vectors, such as malaria-bearing mosquitos [47], with an important role in seed dispersal, pollination, material and nutrient distribution, and recycle [48].

One of the most troubling problems of the farming industry is insect pests, which affect crop production worldwide. Inhibitors of natural insect pests such as bats, as major predators of arthropods, provide valuable ecosystem benefits for crops cultivation [48, 49]. Indeed, herbivorous arthropods destroy approximately 25–50% of crops worldwide [50, 51], in response to these threats, it is clear that the application of synthetic pesticides has increased which in turn has led to several unintended consequences including human health risks, degradation of ecosystem function, evolved toxicity resistance by pests, and severe alterations in the agribusiness dynamics [50, 52, 53].

Indeed, the elimination of beneficial vertebrate predators that act as biocontrol of insects, such as bats, insect species that are not normally considered pests are often elevated to pest status [53, 54]. Thus, the promotion of biological controls can reduce the widespread and indiscriminate use of chemical pesticides [55]. About 99% of potential crop pests are limited by natural ecosystems [53, 56] of which some fraction can be attributed to predation by bats [57]. For example, a colony of 150 big brown bats (*Eptesicus fuscus*) in the midwestern United States annually consume approximately 600,000 spotted cucumber beetles (*Diabrotica undecimpunctata*), 194,000 June beetles (Scarabidae), 158,000 leafhoppers (Cicadellidae), and 335,000 stinkbugs (Pentatomidae), which are severe crop pests [58]. A Brazilian free-tailed bat (*Tadarida brasiliensis*) can eat up to 20 females of corn earworm moth (*Helicoverpa zea*), one of the costliest agricultural pest insects in one night [59]. In fact, each moth can potentially produce 10,000 crop-damaging caterpillars each night [60]. Bats are just one of several groups of animals that naturally prey on mosquitoes. A Florida colony of 30,000 southeastern myotis (*Myotis austroriparius*) eats 50 tons of insects annually, including more than 15 tons of mosquitoes (**Figure 1**) [61, 62].

Therefore, insectivorous bat species, which largely feed on airborne insects and other arthropods, considerably contribute to the suppression of insects that harm the agricultural industry or transmit specific pathogens to humans, consequently contributing to the maintenance of ecosystem stability.

Figure 1.
Insectivorous bat species (a) big brown bats (Eptesicus fuscus); (b) Southeastern myotis (Myotis austroriparius); (c) Brazilian free-tailed bat (Tadarida brasiliensis): photos by Merlin Tuttle's Bat Conservation/MerlinTuttle.org.

Figure 2.
A lesser long-nosed bat (Leptonycteris yerbabuenae) feeding on a cross-sectioned saguaro cactus flower: photo by Merlin Tuttle's Bat Conservation/MerlinTuttle.org.

Another important ecosystem service provided by bats is pollination. Although, bat pollination is relatively uncommon when compared with bird or insect pollination, it involves an impressive number of economically and ecologically important plants [63]. Plant-visiting bats play a remarkable role in facilitating reproductive success and the recruitment of new seedlings [48]. For instance, bat-pollinated columnar cacti (Cactaceae) and agaves (Asparagaceae) are among the most important species as dominant vegetation elements in arid and semiarid habitats (**Figure 2**) [48].

Furthermore, seed dispersal is a major way in which animals contribute to ecosystem succession by depositing seeds from one area to another [64]. Frugivorous bats play a tremendous role in dispersing the seeds of tropical trees and shrubs to produce fleshy fruits as 50–90% of these trees are adapted to the consumption of vertebrates [65]. Generally, frugivorous bats help maintain the diversity of forests by dispersing seeds across different ecosystems, often introducing novel plant species into previously disturbed landscapes [64].

In contrast to predation, which is an antagonistic population interaction, pollination, and seed dispersal are mutualistic population interactions in which plants provide a nutritional reward (nectar, pollen, and fruit pulp) for a beneficial service [48]. Indeed, frugivorous bats as well as nectarivorous bats provide valuable ecosystem services by pollinating plants, dispersing pollen, and thus helping to maintain the genetic diversity of flowering plants.

Bats offer an important multisensory role in assessing ecosystem health either by directly contributing to regulating services to agricultural production or indirectly by providing forage and nesting habitat for pollinators and seed dispersers. To some extent, bat guano has great ecological potential for soil fertility and nutrient distribution, as bats sprinkle them over the landscape overnight, they facilitate nutrient redistribution within ecosystems [66]. In other words, bat guano supports a great diversity of organisms including arthropods, fungi, bacteria, and lichens representing different trophic levels by supporting their ecosystems [67].

Moreover, bats have enormous potential as bioindicators of both disturbance and the existence of contaminants [68]. They show taxonomic stability, trends in their populations can be monitored, short- and long-term effects on populations can be measured and widely distributed worldwide [68–70]. Since insectivorous bats occupy high trophic levels, they are sensitive to the accumulations of pesticides and other toxins, and changes in their abundance may reflect changes in populations of arthropod prey species [71]. In particular, changes in bat numbers or activity can be related to climate change, deterioration of water quality, agricultural intensification, loss and fragmentation of habitats, fatalities at wind turbines, disease, pesticide use, and overhunting [68], and hence bat populations are affected by a wide range of stressors that affect many other taxa. Overall, there is an urgent need to implement a global effort for monitoring bat populations, to ensure that their role as bioindicators can be used to their full potential.

2.2.2 Economic services

Estimating the economic importance of bats in agricultural systems is challenging; however, the bats value in pest suppression illustrates an important agricultural service by increasing the monetary gain of farmers, and consequently supporting food security [49]. They consume enormous quantities of insect pests that cost farmers and foresters billions of dollars annually [59]. In USA, due to the fact of insect pests are eliminated by bat predation, the estimated value of bats as a result of reduced costs of pesticide applications is in the range of \$3.7–\$53 billion per year [66]. These estimates include the reduced costs of unnecessary pesticide applications to suppress insects consumed by bats [72].

Bats provide substantial ecosystem services worldwide, and their benefits to human economies are not limited to agricultural pest control. For example, pioneering research in tropical ecosystems shows the importance of plant-visiting bats in the pollination of valuable fruit crops [73, 74]. There are 289 Old World tropical plant species, which rely on pollination and seed dispersal services by bats for their propagation, providing human with about 448 products in a variety of categories, for instance, wood products (23%); food, drinks and fresh fruit (19%); medicine (15%); dye, fiber, animal fodder, fuel wood, and others (43%) [57]. In addition to some cash crops such as wild bananas (*Musa acuminata*), mangos (*Mangifera indica*), breadfruits (*Artocarpus altilis*), agave (*Agave* spp.), and durians (*Durio* spp.) they rely on bats for pollination [57]. Durian, a wildly popular fruit worth more than \$230 million per year in Southeast Asia, opens its flower at dusk and relies almost exclusively on fruit bats for pollination [57]. In general, according to the vital role that bats play in the global pollination services, their total economic value is up to \$200 billion [75]. Furthermore, bats have a tremendous economic value in maintaining forests through dispersing the seeds of crucial plants for forest reemerging [76]. For instance, the estimated economic value of bat seed dispersal services to giant oak (*Quercus virginiana*) is \$212,000 for acorn seeding and \$945,000 for planting saplings [77].

Bats provide some of the world's finest natural fertilizers known as Guano [78]. Since there are high concentrations of nitrogen and phosphorous in guano [79], it provides some of the world's finest natural fertilizers [78]. For instance, in Texas the

Brazilian free-tailed bat guano has been extracted for fertilizer in thousands of tons from Bracken Cave alone with the current retail sales ranging from $2.86 to $12.10 per kilogram [78].

Finally, it is important to recognize the extraordinary value of bats to ancient and contemporary traditions and science. Recently, bats provide esthetic contributions through cave visits, nocturnal tours in national parks, and educational nature programs. Like other wildlife recreational activities, bat watching is considerably growing [80]. Besides providing adventure and life memories to the public, it generates income for the communities and companies involved [80]. For instance, Congress Avenue Bridge, one of the largest urban bat colonies in the USA, is visited by 200–1500 visitors every evening with a value of $3 million per year [80, 81]. Bats also commonly appear as symbols or logos in popular movies (e.g., Batman), products (e.g., Bacardi rum), and holidays (e.g., Halloween), and all major revenue-generating endeavors [82]. Moreover, many novel technological advances have been inspired by bat echolocation and locomotion, such as sonar systems, biomedical ultrasound, sensors for autonomous systems, wireless communication, and BATMAVs (bat-like motorized aerial vehicles) [83]. Also, in the medical sector, the saliva of vampire bats with its anticlotting chemicals has been investigated as a potential anticoagulant for people who are at high risk of blood clots and strokes [84].

Although, some of these services provide direct benefits to humans (e.g., food, fuel, fiber, and fertilizer), most ecosystem services offer indirect benefits (e.g., pest suppression, seed dispersal, and pollination). Often, little attention is paid to "free" services provided by ecosystems either because the benefits of the services are not fully understood by decision makers or because the benefits accrue to non-owners of the ecosystem providing the service [48]. Information on nonmarket values of ecosystem services can be used to inform decisions regarding whether to protect existing ecosystem services, improve the current provision of ecosystem services, or restore previously lost ecosystem services [4, 85].

Finally, although much of the public and some policy makers may view the precipitous decline of bats worldwide as only of academic interest, the economic and ecological consequences of losing so many bats could be substantial. Thus, a deeper understanding

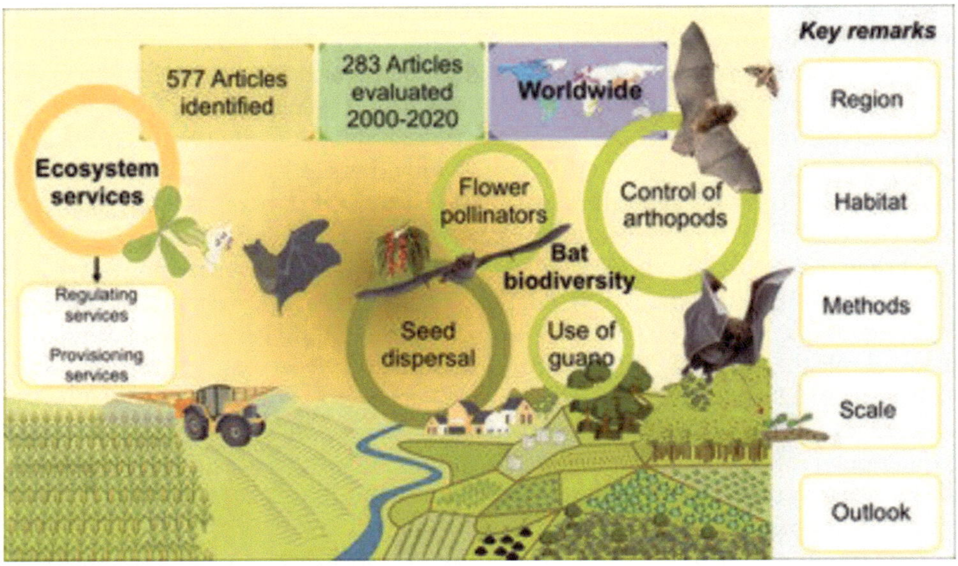

Figure 3.
Ecosystem services provided by bats by Ramírez-Fráncel [86].

of bat-human inter-relationships, the importance of bat diversity, and ecosystem management is crucial for healthy ecosystems and human well-being (**Figure 3**).

3. Ecosystem management

The management and protection of ecosystems are essential to the functioning of ecosphere processes and for the well-being of the biotic and abiotic components of the Earth. Different sectors of society view ecosystem management in terms of their own economic, cultural and societal needs. Management should involve all stakeholders and balance local interests with the broader public interest. Effective ecosystem management depends on both cultural and biological diversity, the dynamic relationship within a species, among species and between species and their abiotic environment, as well as the physical and chemical interactions within the environment [87].

There are different perspectives of environmental management, one of which deals with human needs as central such as protecting a young forest because it may be able to be logged in the future which is known as anthropocentric perspective. While the egocentric perspective deals with the needs of environments as central, for example allowing farmers to extract less water from the river to permit more water to flow downstream [88]. Thus, we must appropriately value and manage ecosystems because of their multiple values to humanity and/or because of their value to other ecosystems and both scenarios lead to a healthier environment and more human well-being.

As human numbers increase and people encroach deeper into the remaining natural habitats, human-bat interactions are becoming more frequent, with often undesirable consequences for both humans and bats [14]. However, bats are often considered keystone species, as they play an important role in many ecosystem services [89], bat populations are declining worldwide mainly because of habitat destruction [90] and increased population control [91, 92]. Additionally, the lack of knowledge about bats makes them an easy target for disease-related fears [93–95] and a potential target for persecution [28, 96].

Indeed, bat management requires a comprehensive approach that must consider the development of culturally appropriate strategies that minimize zoonotic health risks and support bat diversity and its associated ecosystem services [97]. However, even with some communities expressing positive attitudes toward bats, bat control efforts and roost destruction are significant threats to the taxon. Thus, healthy comprehensive management depends on a robust understanding of the importance of bat diversity and emphasizes the non-lethal actions by the general public, health officials, and pest/wildlife managers [98]. It is essential to document people's relationship with bats and to incorporate these perceptions into educational efforts and management decisions, consequently increasing the potential for successful conservation efforts. While comprehensive management can be more costly in the short term (compared to typical pest control efforts), the long-term results should provide the best sustainable outcomes that are satisfactory to people, bats, and the environment [99].

The key to sustainable development is to strike a balance between the exploitation of natural resources for socio-economic development and the preservation of ecosystem services. In other words, healthy ecosystems are a prerequisite for sustainable development and all ecosystem services provided by different vertebrate animals, such as bats, compromise options for present and future generations. Hence, it is important to demonstrate that undervaluing one ecosystem service is likely to lead to the loss of many.

4. Conclusion

In conclusion, for the benefit of present and future generations, bat management by governmental agencies and NGOs needs to be evaluated. Additionally, more understanding is needed of how ecosystem services provided by bats contribute to livelihoods and how far the benefits provided by bats elevate their detrimental impact. Furthermore, in the context of accelerating environmental change, there is an urgent need to identify ecosystem conservation, restoration, and management strategies that are likely to support biodiverse and adaptive ecosystems in the future. We hope that increased awareness about these nocturnal, fast-flying, and secretive mammals can help build synergies between international scientific knowledge, conservation priorities, and local cultural values, which together can promote the benefits of the ecosystem services provided by bats (Video 1, https://www.merlintuttle.org/video/the-importance-of-bats/).

Acknowledgements

We would like to express our special thanks of gratitude to Dr. Merlin D. Tuttle "Founder & Executive Director, Merlin Tuttle's Bat Conservation" and Dr. Ramírez-Fráncel for their worthwhile support to our work.

Author details

Kareem M. Soliman* and Wiame W. Emam
Faculty of Science, Ain Shams University, Cairo, Egypt

*Address all correspondence to: dr_kariemsoliman@yahoo.com

References

[1] Chivian E, Bernstein A, editors. Sustaining life. In: How Human Health Depends on Biodiversity. New York: Oxford University Press; 2008

[2] Daily GC, editor. Nature's Services. Societal Dependence on Natural Ecosystems. Washington, DC: Island Press; 1997

[3] Daily GC, Söderqvist T, Aniyar S, Arrow K, Dasgupta P, Ehrlich PR, et al. The value of nature and the nature of value. Science. 2000;**289**(5478):395-396. DOI: 10.1126/science.289.5478.395

[4] Joseph A, Neville JA, Colin DB, Doris C, Stephen RC, Juan CC, et al. Millennium ecosystem assessment. In: Ecosystems and Human Well-Being: A Framework for Assessment. Washington, DC: Island Press; 2003

[5] Walter VR, Harold A M, Angela C, Doris C, Stephen RC, Kanchan C, et al. Millennium ecosystem assessment. In: Ecosystems and Human Well-Being: Synthesis. Washington, DC: Island Press; 2005

[6] Simmons NB, Seymour KL, Habersetzer J, Gunnell GF. Primitive early eocene bat from Wyoming and the evolution of flight and echolocation. Nature. 2008;**451**:U818-U816. DOI: 10.1038/nature06549

[7] Simmons NB, Cirranello AL. Bat Species of the World: A Taxonomic and Geographic Database [web page]. 2021. Available from: https://batnames.org/

[8] Patterson BD, Willig MR, Stevens RD. Trophic strategies, niche partitioning, and patterns of ecological organization. In: Kunz TH, Fenton MB, editors. Bat Ecology. Chicago: University of Chicago Press; 2003. pp. 536-579

[9] Simmons NB, Conway M. Evolution and ecological diversity of bats. In: Kunz TH, Fenton MB, editors. Bat Ecology. Chicago: University of Chicago Press; 2003. pp. 493-535

[10] Kunz TH. Ecology of Bats. 1st ed. Vol. XVIII. Boston, MA: Springer; 1982. p. 450. DOI: 10.1007/978-1-4613-3421-7

[11] Chaverri G, Kunz TH. Ecological determinants of social systems: Perspectives on the role of roosting ecology in the social behavior of tent-roosting bats. In: Macedo R, editor. Behavioral Ecology of Tropical Animals, Advances in the Study of Behavior. Vol. 42. New York: Academic Press; 2010. pp. 275-318. DOI: 10.1016/S0065-3454(10)42009-4

[12] Lawrence EA. The sacred bee, the filthy pig, and the bat out of hell: Animal symbolism as cognitive biophilia. In: Kellert SR, editor. The Biophilia Hypothesis. Washington, D.C: Island Press; 1993. pp. 301-341

[13] Low M, Hoong WZ, Shen Z, Murugavel B, Mariner N, Paguntalan LM, et al. Bane or blessing? Reviewing cultural values of bats across the Asia-Pacific region. Journal of Ethnobiology. 2021;**41**(1):18-34. DOI: 10.2993/0278-0771-41.1.18

[14] Rocha R, Fernández-Llamazares A, López-Baucells A, Andrimatandrina SFM, Andriatafika ZE, Burgas D, et al. Human-bat interactions in rural southwestern Madagascar through a biocultural lens. Journal of Ethnobiology. 2021;**41**(1):53-69. DOI: 10.2993/0278-0771-41.1.53

[15] Morcote-Ríos G, Aceituno FJ, Iriarte J, Robinson M, Chaparro-Cárdenas JL. colonisation and early peopling of the colombian amazon during the late pleistocene and the early holocene: New evidence from La Serranía La Lindosa. Quaternary International. 2021;**578**:5-19. DOI: 10.1016/j.quaint.2020.04.026

[16] Polák J, Rádlová S, Janovcová M, Flegr J, Landová E, Frynta D. Scary and nasty beasts: Self-reported fear and disgust of common phobic animals. British Journal of Psychology. 2020; **111**(2):297-321. DOI: 10.1111/bjop.12409

[17] Todd J. Dreaming the bat out of the shadow. Psychological Perspectives. 2016;**59**(2):219-241. DOI: 10.1080/00332925.2016.1170566

[18] Deshpande K, Kelkar N. How do fruit bat seed shadows benefit agroforestry? Insights from local perceptions in Kerala, India. Biotropica. 2015;**47**(6):654-659. DOI: 10.1111/btp.12275

[19] Fagan KE, Willcox EV, Willcox AS. Public attitudes toward the presence and management of bats roosting in buildings in Great Smoky Mountains National Park, Southeastern United State. Biological Conservation. 2018; **220**(1):132-139. DOI: 10.1016/j.biocon.2018.02.004

[20] Lim VC, Wilson JJ. Public perceptions and knowledge of, and responses to, bats in urban areas in Peninsular Malaysia. Anthrozoös. 2019;**32**(6):825-834. DOI: 10.1080/08927936.2019.1673063

[21] Shapiro HG, Willcox AS, Ader DR, Willcox EV. Attitudes towards and relationships with cave-roosting bats in northwest Cambodia. Journal of Ethnobiology. 2021;**41**(1):87-104. DOI: 10.2993/0278-0771-41.1.87

[22] Fernández-Llamazares Á, Cabeza M. Rediscovering the potential applications of indigenous storytelling for conservation practice. Conservation Letters. 2018;**11**(3):e12398. DOI: 10.1111/conl.12398

[23] Kingston T. Cute, creepy, or crispy—How values, attitudes and norms shape behavior toward bats. In: Voigt CC, Kingston T, editors. Bats in the Anthropocene: Conservation of Bats in a Changing World. Springer International Publishing; 2015. pp. 571-595. DOI: 10.1007/978-3-319-25220-9_18

[24] Lunney D, Moon C. Blind to bats. In: Law B, Eby P, Lunney D, Lumsden L, editors. The Biology and Conservation of Australasian Bats. Mosman, NSW, Australia: Royal Zoological Society of NSW; 2011. pp. 44-63. DOI: 10.7882/FS.2011.008

[25] Musila S, Prokop P, Gichuki N. Knowledge and perceptions of, and attitudes to, bats by people living around Arabuko-Sokoke Forest, Malindi-Kenya. Anthrozoös. 2018;**31**: 247-262. DOI: 10.1080/08927936.2018.1434065

[26] Tuttle MD. Fear of bats and its consequences. Journal of Bat Research and Conservation. 2017;**10**(1):66-69. DOI: 10.14709/BarbJ.10.1.2017.09

[27] Prokop R, Tunnicliffe SD. "Disgusting" animals: Primary school children's attitudes and myths of bats and spiders. Eurasia Journal of Mathematics, Science & Technology Education. 2008;**4**(2):87-97. DOI: 10.12973/ejmste/75309

[28] Prokop P, Fančovičová J, Kubiatko M. Vampires are still alive: Slovakian students' attitudes toward bats. Anthrozoös. 2009;**22**:19-30. DOI: 10.2752/175303708X390446

[29] Rydell J, Eklöf J, Riccucci M, du Père-Lachaise C. Bats and vampires in French romanticism. Journal of Bat Research & Conservation. 2018;**11**:83-91. DOI: 10.14709/BarbJ.11.1.2018.10

[30] Tanalgo KC. Wildlife hunting by indigenous people in a Philippine protected area: A perspective from Mt. Apo National Park, Mindanao Island. Journal of Threatened Taxa. 2017;**9**: 10307-10313. DOI: 10.11609/jott.2967.9.6.10307-10313

[31] Tanalgo KC, Teves RD, Salvaña FRP, Baleva RE, Tabora JAG. Human-bat interactions in caves of south-central Mindanao, Philippines. Wildlife Biology in Practice. 2016;**12**:1-14

[32] Tatai E. An iconographical approach to representations of the devil in medieval Hungary. Christian Demonology and Popular Mythology. 2006;**2**:54

[33] Charro M. Murciélagos: Príncipes de las Tinieblas. Revista Folklore. 1999;**19**:111-118

[34] Frembgen JW. Embodying evil and bad luck: Stray notes on the folklore of bats in southwest Asia. Asian Folklore Studies. 2006;**65**(2):241-247. Available from: https://www.jstor.org/stable/30030400

[35] Eklöf J, Rydell J. Attitudes towards bats in swedish history. Journal of Ethnobiology. 2021;**41**:35-52. DOI: 10.2993/0278-0771-41.1.35

[36] Pham HMA, Yonn-Soon L. A study on the classifications and symbolic meanings of Vietnamese traditional patterns. International Journal of Human Ecology. 2008;**9**(1):29-40

[37] Hall LS, Richards G. Flying Foxes: Fruit and Blossom Bats of Australia. Sydney: UNSW Press; 2000

[38] Hadnes M, Schumacher H. The Gods are watching: An experimental study of religion and traditional belief in Burkina Faso. Journal for the Scientific Study of Religion. 2012; **51**(4):689-704. DOI: 10.1111/j.1468-5906.2012.01676.x

[39] Retana-Guiascón OG, Navarijo-Ornelas ML. Los valores culturales de los murciélagos. Revista Mexicana de Mastozoología. 2012;**2**(1): 18-26. DOI: 10.22201/ie.20074484e.2012.2.1.19

[40] Voigt CC, Phelps KL, Aguirre LF, Corrie Schoeman M, Vanitharani J, Zubaid A. Bats and buildings: The conservation of synanthropic bats. In: Voigt CC, Kingston T, editors. Bats in the Anthropocene: Conservation of Bats in a Changing World. Cham, Switzerland: Springer; 2016. pp. 427-462. DOI: 10.1007/978-3-319-25220-9_14

[41] Aziz SA, Clements GR, Giam X, Forget PM, Campos-Arceiz A. Coexistence and conflict between the island flying fox (*Pteropus hypomelanus*) and humans on Tioman Island, Peninsular Malaysia. Human Ecology. 2017;**45**:377-389. DOI: 10.1007/s10745-017-9905-6

[42] The Concept of the Ecosystem. Umich. Edu. 2017. Available from: https://globalchange.umich.edu/globalchange1/current/lectures/kling/ecosystem/ecosystem.html

[43] Maas B, Karp DS, Bumrungsri S, Darras K, Gonthier D, Huang JC, et al. Bird and bat predation services in tropical forests and agroforestry landscapes. Biological Reviews. 2016;**91**:1081-1101. DOI: 10.1111/brv.12211

[44] Puig-Montserrat X, Torre I, López-Baucells A, Guerrieri E, Monti MM, Ràfols-García R, et al. Pest control service provided by bats in mediterranean rice paddies: Linking agroecosystems structure to ecological functions. Mammalian Biology. 2015;**80**:237-245. DOI: 10.1016/j.mambio.2015.03.008

[45] Puig-Montserrat X, Flaquer C, Gómez-Aguilera N, Burgas A, Mas M, Tuneu C, et al. Bats actively prey on mosquitoes and other deleterious insects in rice paddies: Potential impact on human health and agriculture. Pest Management Science. 2020;**76**:3759-3769. DOI: 10.1002/ps.5925

[46] Wanger TC, Darras K, Bumrungsri S, Tscharntke T, Klein AM. Bat pest control contributes to food security in Thailand. Biological

Conservation. 2014;**171**:220-223. DOI: 10.1016/j.biocon.2014.01.030

[47] Kemp J, López-Baucells A, Rocha R, Wangensteen OS, Andriatafika Z, Nair A, et al. bats as potential suppressors of multiple agricultural pests: A case study from Madagascar. Agriculture, Ecosystems & Environment. 2019;**269**: 88-96. DOI: 10.1016/j.agee.2018.09.027

[48] Kunz TH, de Torrez EB, Bauer D, Lobova T, Fleming TH. Ecosystem services provided by bats. Annals of the New York Academy of Sciences. 2011;**1223**:1-38. DOI: 10.1111/j.1749-6632.2011.06004.x

[49] de Groot R, Brander L, van der Ploeg S, Costanza R, Bernard F, Braat L, et al. Global estimates of the value of ecosystems and their services in monetary units. Ecosystem Services. 2012;**1**:50-61. DOI: 10.1016/j.ecoser.2012.07.005

[50] Pimentel D, Krummel J, Gallahan D, Hough J, Merrill A, Schreiner I, et al. Benefits and costs of pesticide use in the US food production. Bioscience. 1978;**28**:778-784. DOI: 10.2307/1307251

[51] Pimentel D, McLaughlin L, Zepp A, Lakitan B, Kraus T, Kleinman P, et al. Environmental and economic effects of reducing pesticide use. Agriculture, Ecosystems & Environment. 1993; **46**:1-4. DOI: 10.1016/0167-8809(93)90030-S

[52] Benbrook CM. Pest management at the crossroads. New York: Consumers Union (1996). The Journal of Agricultural Science. 1998;**130**:119-123. DOI: 10.1017/S0021859697245120

[53] Naylor RL, Ehrlich PR. Natural pest control services and agriculture. In: Daily GC, editor. Nature's Services. Societal Dependence on Natural Ecosystems. Washington, DC: Island Press; 1997. pp. 151-176

[54] National Research Council. Alternative Agriculture. Washington, DC: National Academies Press; 1989

[55] United States Environmental Protection Agency. Integrated Pest Management (IPM) Principles. 2009. Available from: http://www.epa.gov/opp00001/factsheets/ipm.htm

[56] Debach P, Rosen D. Biological Control by Natural Enemies. 2nd ed. London: Cambridge University Press; 1991. SB975.D39 1991

[57] Fujita MS, Tuttle MD. Flying foxes (Chiroptera: Pteropodidae): Threatened animals of key ecological and economic importance. Conservation Biology. 1991;**5**:455-463. DOI: 10.1111/j.1523-1739.1991.tb00352.x

[58] Whitaker JO. Food of the big brown bat Eptesicus fuscus from maternity colonies in Indiana and Illinois. American Midland Naturalist. 1995;**134**: 346-360. DOI: 10.2307/2426304

[59] Keeley BW, Tuttle MD. Bats in American Bridges. Resource Publication #4. Austin, Tex.: Bat Conservation International, Inc.; 1999

[60] McCracken GF. Bats aloft: A study of high-altitude feeding. Bats. 1996;**14**:7-101

[61] Anthony ELP, Kunz TH. Feeding strategies of the little brown bat (*Myotis lucifugus*) in southern New Hampshire. Ecology. 1977;**58**:775-786. DOI: 10.2307/1936213

[62] Reiskind MH, Wund MA. Experimental assessment of the impacts of northern long-eared bats on ovipositing Culex (Diptera: Culicidae) mosquitoes. Journal of Medical Entomology. 2009;**46**:1037-1044. DOI: 10.1603/033.046.0510

[63] Fleming TH, Muchhala N. Nectar-feeding bird and bat niches in two

worlds: Pantropical comparisons of vertebrate pollination systems. Journal of Biogeography. 2008;**35**:764-780. DOI: 10.1111/j.1365-2699.2007.01833.x

[64] Duncan RS, Chapman CA. Seed dispersal and potential forest succession in abandoned agriculture in tropical Africa. Ecological Applications. 1999;**9**:998-1008. DOI: 10.1890/1051-0761

[65] Howe HF, Smallwood J. Ecology of seed dispersal. Annual Review of Ecology and Systematics. 1982;**13**:201-228. DOI: 10.1146/annurev.es.13.110182.001221

[66] Boyles JG, Cryan PM, McCracken GF, Kunz TH. Economic importance of bats in agriculture. Science. 2011;**332**:41-42. DOI: 10.1126/science.1201366

[67] Polis GA, Anderson WB, Holt RD. Toward an integration of landscape and food web ecology: The dynamics of spatially subsidized food webs. Annual Review of Ecology and Systematics. 1997;**28**:289-316. DOI: 10.1146/annurev.ecolsys.28.1.289

[68] Jones G, Jacobs D, Kunz TH, Wilig MR, Racey PA. CarpeNoctem: The importance of bats as bioindicators. Endangered Species Research. 2009;**8**:93-115. DOI: 10.3354/esr00182

[69] Fenton MB. Science and the conservation of bats: Where to next? Wildlife Society Bulletin. 2003;**31**:6-15. DOI: 10.2307/3784355

[70] Fenton MB, Acharya L, Audet D, Hickey MBC, Merriman C, Obrist MK, et al. Phyllostomid bats (Chiroptera: Phyllostomidae) as indicators of habitat disruption in the neotropics. Biotropica. 1992;**24**:440-446. DOI: 10.2307/2388615

[71] Hutson AM, Mickleburgh SP, Racey PA. Microchiropteran Bats: Global Status Survey and Conservation Action Plan. Gland, Switzerland: IUCN/SSC chiroptera specialist group, IUCN; 2001. DOI: 10.2305/IUCN.CH.2001.SSC-AP.1.en

[72] Cleveland CJ, Betke M, Federico P, Frank JD, Hallam TG, Horn J, et al. Economic value of the pest control service provided by Brazilian free-tailed bats in south-central Texas. Frontiers in Ecology and the Environment. 2006;**4**:238-243. DOI: 10.1890/1540-9295

[73] Bumrungsri S, Sripaoraya E, Chongsiri T, Sridith K, Racey PA. The pollination ecology of durian (*Durio zibethinus*, Bombacaceae) in southern Thailand. Journal of Tropical Ecology. 2009;**25**:85-92. DOI: 10.1017/S0266467408005531

[74] Bumrungsri S, Harbit A, Benzie C, Carmouche K, Sridith K, Racey PA. The pollination ecology of two species of Parkia (Mimosaceae) in southern Thailand. Journal of Tropical Ecology. 2008;**24**:467-475. DOI: 10.1017/S0266467408005191

[75] Gallai N, Salles JM, Settele J, Vaissi'ere BE. Economic valuation of the vulnerability of world agriculture confronted with pollinator decline. Ecological Economics. 2009;**68**:810-821. DOI: 10.1016/j.ecolecon.2008.06.014

[76] Buddenhagen IW. Bats and disappearing wild bananas: Can bats keep commercial bananas on supermarket shelves? Bats. 2008;**26**:1-6

[77] Hougner C, Colding J, Söderqvist T. Economic valuation of a seed dispersal service in the Stockholm National Urban Park, Sweden. Ecological Economics. 2006;**59**:364-374. DOI: 10.1016/j.ecolecon.2005.11.007

[78] Tuttle MD, Moreno A. Cave-Dwelling Bats of Northern Mexico: Their Value and Conservation Needs.

Austin, Texas, USA: Bat Conservation International, Inc.; 2005

[79] Hutchinson GE. Survey of existing knowledge of biogeochemistry: The biogeochemistry of vertebrate excretion. Bulletin of the American Museum of Natural History. 1950;**96**:1-554. Available from: http://hdl.handle.net/2246/918

[80] Tapper R. Wildlife Watching and Tourism: A Study on the Benefits and Risks of a Fast Growing Tourism Activity and Its Impacts on Species. Bonn, Germany: UNEP/CMS Secretariat; 2006

[81] Pennisi LA, Holland SM, Stein TV. Achieving bat conservation through tourism. Journal of Ecotourism. 2004;**3**:195-207. DOI: 10.1080/14664200508668432

[82] Brown DE. Vampiro: The Vampire Bat in Fact and Fantasy. 1st ed. Silver City, NM: High-Lonesome Books; 1994

[83] Bunget G, Seelecke S. BATMAV: A 2-DOF bio-inspired flapping flight platform. The International Society for Optics and Photonics. 2010;**7643B**:1-11. DOI: 10.1117/12.853398

[84] Schleuning WD. Vampire bat plasminogen activator DSPAalpha-1 (desmoteplase): A thrombolytic drug optimized by natural selection. Pathophysiology of Haemostasis and Thrombosis. 2000;**31**:118-122. DOI: 10.1159/000048054

[85] National Research Council. Valuing Ecosystem Services: Towards Better Environmental Decision-Making. Washington, DC: National Academies Press; 2005

[86] Ramírez-Fráncel LA, García-Herrera LV, Losada-Prado S, Reinoso-Flórez G, Sánchez-Hernández A, Estrada-Villegas S, et al. Bats and their vital ecosystem services: A global review. Integrative Zoology. 2021;**0**:1-22. DOI: 10.1111/1749-4877.12552

[87] Bellera EE, McClenachan L, Zavaleta ES, Larsen LG. Past forward: Recommendations from historical ecology for ecosystem management. Global Ecology and Conservation. 2020;**21**:e00836. DOI: 10.1016/j.gecco.2019.e00836

[88] Rülke J, Rieckmann M, Nzau JM, Teucher M. How ecocentrism and anthropocentrism influence human–environment relationships in a Kenyan biodiversity hotspot. Sustainability. 2020;**12**:8213. DOI: 10.3390/su12198213

[89] Ghanem SJ, Voigt CC. Increasing awareness of ecosystem services provided by bats. Advances in the Study of Behavior. 2012;**44**:279-302. DOI: 10.1016/b978-0-12-394288-3.00007-1

[90] Elliot WR. Protecting caves and cave life. In: Culver DC, White WB, editors. Encyclopedia of Caves. Amsterdam: Elsevier; 2014. pp. 458-467

[91] Reid JL. Knowledge and experience predict indiscriminate bat-killing intentions among Costa Rican men. Biotropica. 2016;**48**:394-404. DOI: 10.1111/btp.12279

[92] Frick WF, Kingston T, Flanders J. A review of the major threats and challenges to global bat conservation. Annals of the New York Academy of Sciences. 2020;**1469**:5-25. DOI: 10.1111/nyas.14045

[93] Olival KJ, Hosseini PR, Zambrana-Torrelio C, Ross N, Bogich TL, Daszak P. Host and viral traits predict zoonotic spillover from mammals. Nature. 2017;**546**:646-650. DOI: 10.1038/nature22975

[94] López-Baucells A, Rocha JR, Fernández-Llamazares Á. When bats go viral: Negative framings in virological

research imperil bat conservation. Mammal Review. 2018;**48**:62-66. DOI: 10.17863/CAM.21045

[95] Centers for Disease Control and Prevention (CDC). Coronavirus disease 2019 (COVID-19). Atlanta, Georgia, USA: Centers for Disease Control and Prevention; 2020. Available from: https://www.cdc.gov/coronavirus/2019-ncov/cases-updates/summary.html#emergence

[96] Dickman AJ, Hazzah L. Money, myths and man-eaters: Complexities of human– wildlife conflict. In: Angelici FM, editor. Problematic wildlife. Springer; 2016. pp. 339-356. DOI: 10.1007/978-3-319-22246-2_16

[97] MacFarlane D, Rocha R. Guidelines for communicating about bats to prevent persecution in the time of COVID-19. Biological Conservation. 2020;**248**:108650. DOI: 10.1016/j.biocon.2020.108650

[98] Fagan KE, Willcox EV, Willcox AS. Public attitudes toward the presence and management of bats roosting in buildings in Great Smoky Mountains National Park, Southeastern United States. Biological Conservation. 2018;**220**:132-139. DOI: 10.1016/j.biocon.2018.02.004

[99] Frantz SC, Laniewicz BR. Comprehensive management of commensal bats. In: Brittingham MC, Kays J, McPeake R, editors. The Ninth Wildlife Damage Management Conference Proceedings. PA, USA: State College; 2000. Available from: http://wildlifedamage.unl.edu

Section 5

Dispersal Patterns and Diversity

Chapter 5

Diversity and Conservation of Bats in Jordan

Zuhair S. Amr, Omar A. Abed and Mohammad Abu Baker

Abstract

The diversity and the conservation status of bats in Jordan are discussed based on recent studies. The bat fauna of Jordan consists of 26 bat species belonging to nine families (Emballonuridae, Hipposideridae, Pteropodidae, Miniopteridae, Molossidae, Nycteridae, Rhinolophidae, Rhinopomatidae, and Vespertilionidae). Bat echolocation calls for some selected species are included. Conservation status based on regional assessment according to the IUCN standards is amended, along with the current legislative laws for the conservation of bats. Threats affecting the bats of Jordan are highlighted including the recent introduction of wind farms and other mining activities. In addition, the role of bats in disease transmission is included.

Keywords: bat, diversity, threats, Jordan

1. Introduction

Jordan is situated at a crossroad between three continents and with diverse habitats (Mediterranean, Saharo-Arabian, Irano-Turanian, and Afro-tropical). Although Jordan is a small country, the bat fauna is diverse with 26 species representing nine families. Within the past four decades, our knowledge of the bats of Jordan expanded significantly, adding new records [1–3], distributional data [4–12], ectoparasites [13], karyotypic studies [14, 15], activity patterns [16], and conservation [17]. Yet, these studies also pointed out a significant shortage in our knowledge especially about the ecology and conservation of the bat fauna of this country. Benda et al. [18] published the most comprehensive and up-to-date manuscript on the bats of Jordan, including distributional data, ecology, echolocation, ectoparasites, and zoogeographical analysis.

This summary on the bats of Jordan is based on cumulative research and field observation since 1978. Over the past four decades, the senior author was involved in numerous studies on various aspects of the mammalian fauna of Jordan, including bats. As a result, several additional records of the bat fauna were added, and further knowledge on habitat preference and threats affecting bats in Jordan was gained. As a developed country, Jordan witnessed accelerated changes in its natural setting (water resources, agricultural practices, urination, mining, etc.), which, in turn, affected the well-being of several species of animals, including bats.

2. Biogeography of Jordan

Jordan is influenced by four major biogeographic regions (**Figure 1**). Vegetation cover, soil texture, altitude, and annual rainfall are among the major factors that shaped these biogeographic regions. Several types of habitats are present in Jordan, ranging from extreme desertic to mild-forested Mediterranean (**Figure 2**). It was agreed on the delineation of these four regions based on vegetation cover as well as animal distribution in Jordan [19, 20].

2.1 The Mediterranean region

This area is represented by mountain ranges extending from Irbid in the north near to the south around Ra's Al Naqb. This region is characterized by the presence of several types of forests (The Phoenicean juniper, *Juniperus phoenicea*, Mt. Atlas mastic tree, *Pistacia atlantica*, Aleppo pine, *Pinus halepensis*, Palestine oak, *Quercus calliprinos*, kermes oak, *Quercus coccifera* and Mount Tabor oak, *Quercus ithaburensis*). Non-forested areas are characterized by a dense cover of the Thorny Burnet, *Sarcopoterium spinosum*. Altitude ranges from 700 to 1500 m asl, with 400–600 mm average annual rainfall and snowfall during winter. Several types of soil are found, including calcareous, *terra rosa*, sandy, and sandy-loamy caused by erosion of the Nubian sandstone that is common in the south [20].

2.2 Irano-Turanian region

This area surrounds the Mediterranean region from the west and the east. It extends to the Syrian Desert to the northeast. Few scattered trees can be observed, mainly *J. phoenicea* and *P. atlantica* trees. Shrubs of *Anabasis articulata*, *Artemesia herba-alba*, *Astragulus spinosum*, *Retama raetam*, *Urginea maritima*, *Ziziphus lotus*,

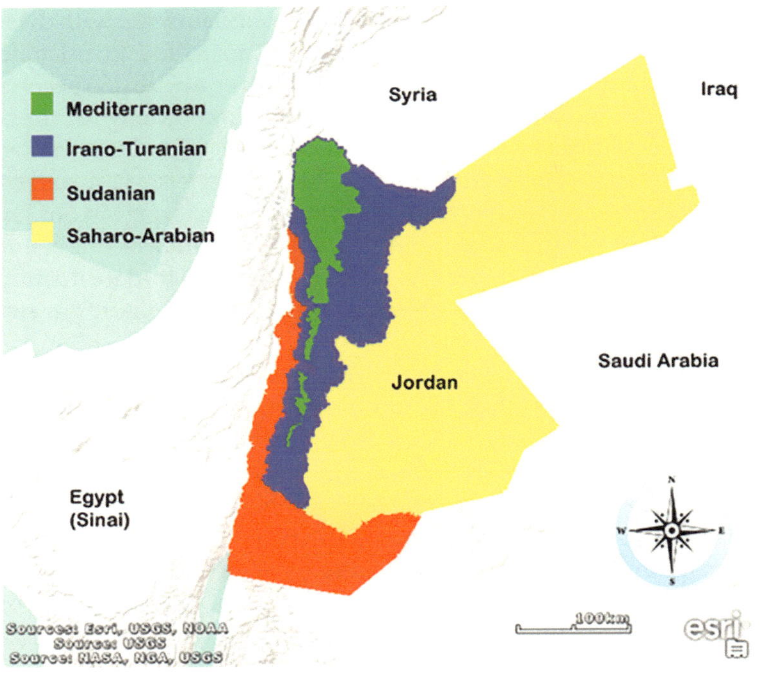

Figure 1.
The biogeographic regions of Jordan (modified after Al-Eisawi [19]).

Figure 2.
Habitat types: (A, B) Temperate Mediterranean habitat with an abundance of evergreen oak (Quercus sp.) and pine forests in northern Jordan. (C) Sand dunes with Haloxylon shrubs and Acacia trees in Wadi Araba. (D) Open Hamada in eastern Jordan with ample bushes of Seidlitzia rosmarinus. (E) Wadi Ramm with sandstone mountains. (F) The riparian habitat at the Yarmouk River Basin in northern Jordan.

Zygophyllum dumosum are dominant. Altitudes range from 400 to 700 m asl, with 50–100 mm average annual rainfall. Surface soil is very thin or absent in some instances, while surface rock outcrops are very high [20].

2.3 Sudanian penetration region

This ecozone extends from southern Jordan near Aqaba Gulf along Wadi Araba reaching the southern part of the Jordan Valley near Dayr Alla northward. It also extends to southeastern Jordan near Wadi Ram, the largest sandstone mountain formation and granite mountains to the east. Altitudes range from 400 m bls around the Dead Sea area to as high as 1734 m asl for Jabal Ram mountain. Precipitation is of less than 50 mm annually. Two species of Acacia occur in this region in varying densities, *Acacia raddiana* and *Acacia tortilis*. *Tamarix* spp. and *Ziziphus spina-christi* are also found along wadi beds. A variety of shrubs, *Anabasis articulata*, *Gymnocarpos decandrum*, *Haloxylon persicum*, and *Lycium* sp. are abundant. The soil is mostly sandy for most of the region, interrupted with some rocky areas [20].

2.4 Saharo-Arabian region

This region constitutes the largest biogeographical region, covering about 70% of Jordan. It spreads to the east bordering the Irano-Turanian region from the west and the Sudanian Penetration region from the southwest. *Haloxylon persicum*, *Hammada scoparia*, and *Ochradenus baccatus* are the typical sand dune vegetation. *Achillea fragrantissima*, *Artemisia herba-alba*, and *Astragalus* sp. are usually associated with wadi beds, and in certain areas east of Ma'an, few scattered *Acacia tortilis* are also found. The soil is mostly gravel, sandy Hamada, saline, and sandy. Altitude ranges from 100 m to 800 m asl, with annual rainfall not exceeding 50 mm. Azraq Oasis, one of the most important nature reserves in Jordan, is located in the middle of the eastern desert [20].

3. Bat diversity in Jordan

The bat fauna of Jordan consists of 26 bat species belonging to nine families (Emballonuridae, Hipposideridae, Pteropodidae, Miniopteridae, Molossidae, Nycteridae, Rhinolophidae, Rhinopomatidae, and Vespertilionidae). **Figure 3** shows some representative species.

3.1 Family Pteropodidae

This family of fruit bats includes a single species, the Egyptian fruit bat, *Rousettus aegyptiacus*. This species is distributed along the eastern mountains and the Jordan Valley, extending from the extreme north near Lake Tiberius reaching as far as Aqaba to the south. The Egyptian fruit bat does not penetrate into the eastern desert [18, 20].

3.2 Family Rhinopomatidae

This family is represented by two species, the greater rat-tailed bat, *Rhinopoma microphyllum* and the lesser mouse-tailed bat, *Rhinopoma cystops*. The later species is widely distributed in localized Mediterranean areas as well as around the Dead Sea basin. *Rhinopoma microphyllum* was reported from fewer localities, mostly in the Mediterranean mountains [18, 20].

3.3 Family Rhinolophidae

This family includes six species inhabiting a variety of habitats. Geoffroy's horseshoe bat, *Rhinolophus clivosus*, and Mehely's horseshoe bat, *Rhinolophus mehelyi* were recorded from arid regions, while the Mediterranean horseshoe bat, *Rhinolophus euryale*, and the larger horseshoe bat, *Rhinolophus ferrumequinum*, are strictly confined to forested areas in northern Jordan. The lesser horseshoe bat, *Rhinolophus hipposideros* and Blasius's horseshoe bat, *Rhinolophus blasii* were recorded from both Mediterranean areas in the north and arid regions in southern Jordan [18, 20].

3.4 Family Emballonuridae

The tomb bat, *Taphozous perforatus* and the naked bellied tomb bat, *Taphozous nudiventris*, were recorded from Jordan. Both species were reported from the middle

DOI: http://dx.doi.org/10.5772/intechopen.100407

Figure 3.
(A) The lesser mouse-tailed bat, Rhinopoma cystops. (B) Geoffroy's horseshoe bat, Rhinolophus clivosus. (C) The desert long-eared bat, Otonycteris hemprichii. (D) Christie's big-eared bat, Plecotus christii. (E) The notch-eared bat, Myotis emarginatus. (F) The trident leaf-nosed bat, Asellia tridens.

part of the Rift Valley of Jordan, surrounding the Dead Sea area. They are found close to open waterbodies such as rivers and creeks [18, 20].

3.5 Family Hipposideridae

The trident leaf-nosed bat, *Asellia tridens*, is an inhabitant of extreme arid areas. It was recorded from several localities in Wadi Araba [18, 20].

3.6 Family Miniopteridae

Miniopterus pallidus is confined to the Mediterranean regions of Jordan forming a large mixed colonies with other bats [18, 20].

3.7 Family Molossidae

The European free-tailed bat, *Tadarida teniotis*, is widely distributed in Jordan. Besides its presence in mild Mediterranean areas, this species can live in extremely dry habitats in the eastern desert of Jordan [18, 20].

3.8 Family Nycteridae

The Egyptian slit-faced bat, *Nycteris thebaica*, is known from the southern parts of Jordan inhabiting barren mountainous areas overlooking the Dead Sea [18, 20].

3.9 Family Vespertilionidae

This is the most diverse family in Jordan and is presented by seven genera (*Barbastella*, *Eptesicus*, *Hypsugo*, *Myotis*, *Otonycteris*, *Pipistrellus*, and *Plecotus*) with 11 species [18, 20].

The Asian barbastelle, *Barbastella leucomelas*, was recorded from the southern regions of Jordan with the semi-arid Mediterranean and extreme desert habitats. It is considered as an endemic species to arid regions around the most northern parts of the Red Sea. Botta's serotine bat, *Eptesicus bottae*, is found in a wide range of semi-arid habitats including the semi-arid Mediterranean regions, as well as low-lands and rocky mountains. This is a crevice-dwelling species, inhabiting buildings, ruins, and natural rock crevices [18, 20].

Christie's big-eared bat, *Plecotus christii*, is a rather common species in Jordan, with distribution limited to the arid regions to the southwestern parts of the country. The desert long-eared bat, *Otonycteris hemprichii* was reported from the northeastern deserts and the arid southwestern parts of Jordan [18, 20].

Kuhl's pipistrelle, *Pipistrellus kuhlii*, is the most common species in Jordan inhabiting all biogeographical regions including the Mediterranean and semi-desert zones. It is very common in urban areas as well. The common pipistrelle, *Pipistrellus pipistrellus*, is distributed along with the western parts of the country in forested areas as well as the arid region. The desert pipistrelle, *Hypsugo ariel*, is known to occur in the arid regions around the Dead Sea basin and the Wadi Ramm desert [18, 20].

The lesser mouse-eared bat, *Myotis blythii*, is restricted to the Mediterranean region of northern Jordan. It forages in scrub and grassland habitats, including farmland and gardens. Maternity colonies are usually found in underground habitats such as caves and old mines. The long-fingered bat, *Myotis capaccinii*, was reported from one locality, Tabqat Fahl, adjacent to the Jordan Valley. The notch-eared bat, *Myotis emarginatus*, was recorded from a limited area of the Ajlun Mountains within the Mediterranean biogeographical region. Natterer's Bat, *Myotis nattereri*, was found in the northwestern forests and from mountainous habitats of the Dhana-Shawbak region in the Southern highlands [18, 20].

4. Bat echolocation calls

Bat echolocation calls for 11 bat species from Jordan are documented. **Figure 4** and **Table 1** show recorded calls and durations and frequency variables using Song meter from different localities.

Species of the genus *Rhinolophus* are characterized by calls with a long and strictly constant-frequency component (**Figure 4**). The four species studied are very distinct; whereas *Rh. ferrumequinum* has the least frequency variables (75.7 ± 4 for SF, and 77.9 ± 6.5 for EF) as compared to *Rh. clivosus*, *Rh. euryale*, and *Rh. hipposideros*

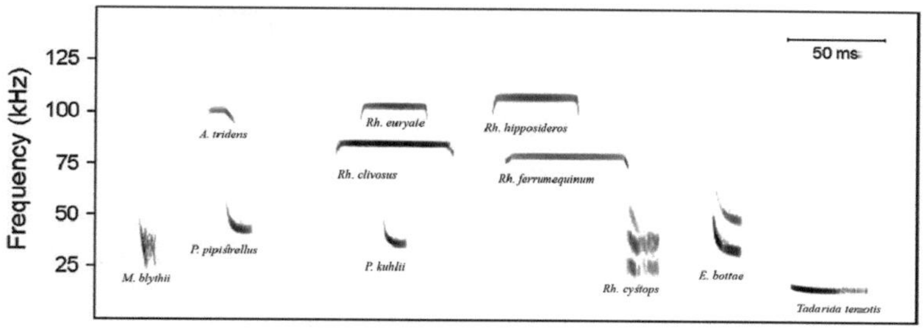

Figure 4.
Representative echolocation calls for 11 species from Jordan.

Species	No. of calls	D [ms]	SF [kHz]	EF [kHz]	PF [kHz]
Rh. clivosus	116	28.7 ± 12.3	85.8 ± 1.7	85.9 ± 1.7	86.2 ± 0.8
Rh. euryale	7	23.4 ± 15.2	105.3 ± 0.9	98.5 ± 7.4	105.8 ± 0.5
Rh. ferrumequinum	107	42 ± 10.5	75.7 ± 4	77.9 ± 6.5	82.2 ± 0.5
Rh. hipposideros	145	29.8 ± 13	107 ± 4.8	107 ± 5.1	111 ± 1.39
Rh. cystops	14	3.6 ± 2.25	35.2 ± 1.9	28.94 ± 0.9	31.2 ± 0.85
A. tridens	103	8.7 ± 3.3	115 ± 2	109 ± 6.2	116 ± 2
E. bottae	105	5.6 ± 2	38.1 ± 4	30.8 ± 1	32.6 ± 1.2
P. kuhlii	76	5.53 ± 1.57	50.5 ± 6.3	39.3 ± 2.2	40.4 ± 2.2
P. pipistrellus	130	5 ± 2	51.7 ± 5	47.9 ± 2.6	48.4 ± 2.4
M. blythii	7	2.3 ± 0.9	46.8 ± 3.5	37.2 ± 3.5	41.8 ± 3
T. teniotis	26	66 ± 1.02	20.2 ± 2.3	15.06 ± 0.12	15.2 ± 0.5

D, call duration; SF, start frequency; EF, end frequency; PF, peak frequency.

Table 1.
Echolocation calls for 11 species recorded from Jordan.

Cave	Type	Number of bat species recorded	Species observed	References
Al Hamma	Karstic	1	*R. aegyptiacus*	[7, 18]
Al Majdal	Karstic	2	*Rh. cystops, Rh. microphyllum*	[11]
Al Wardeh		5	*Rh. blasii, Rh. euryale, M. blythii, M. emarginatus, M. pallidus*	[18]
Al-Mahhatta	Pressure ridge	1	*R. aegyptiacus*	[1]
Ar Raddass		1	*T. nudiventris*	[18]
Arjan		2	*M. emarginatus, P. kuhlii*	[18]
Bir Hamma cave	Lava	1	*O. hemprichii*	[18]
Dhana village caves	Sandstone	2	*E. bottae, P. christii*	[18]
Dibbin Forest caves		4	*Rh. cystops, Rh. euryale, Rh. ferrumequinum, Rh. hipposideros*	[1, 11, 14, 18]

Cave	Type	Number of bat species recorded	Species observed	References
Iraq Al Amir	Artificial	6	*R. aegyptiacus, Rh. cystops, Rh. blasii, M. nattereri, P. kuhlii, T. teniotis*	[18]
Iraq Al Wahaj	Karstic	6	*R. aegyptiacus, Rh. blasii, Rh. euryale, M. blythii, M. emarginatus, M. nattereri*	[18]
Jabal Al Bayda	Sandstone	2	*Rh. clivosus, P. christii*	[18]
Jesus' cave		1	*Rh. hipposideros*	[18]
Lot's cave		1	*Rh. cystops*	[18]
Mahjub cave	Sandstone	3	*Rh. clivosus, M. nattereri, P. christii*	[18]
Malka cave	Artificial	1	*Rh. ferrumequinum*	[18]
Mgharet Issa		1	*Rh. cystops*	[18]
Mogharet Al-Roum Cave		1	*Rh. blasii*	[1]
Mukawir		1	*P. christii*	[18]
Qaraiqira cave		1	*Rh. cystops*	[11]
Tabaqat Fahl	Artificial	5	*R. aegyptiacus, Rh. microphyllum, Rh. blasii, Rh. ferrumequinum, M. capaccinii*	[14, 18]
Umm Al Iraq		1	*Rh. ferrumequinum*	[18]
Wadi Al Hasa		1	*N. thebaica*	[18]
Wadi Ben Hammad	Limestone	1	*R. aegyptiacus*	[20]
Wadi Dhana	Sandstone	1	*R. aegyptiacus*	[18]
Wadi Dharih	Karstic	1	*Rh. clivosus*	[18]
Wadi Shu'ayb		1	*Rh. cystops*	[18]
Zubiya	Karstic	6	*R. blasii, Rh. euryale, Rh. ferrumequinum, Rh. hipposideros, M. emarginatus, M. nattereri*	[18]

Table 2.
Important bat caves with recorded species.

(**Table 1**). *Rhinopoma cystops* exhibits two harmonics; the first around 33 kHz and the second at 32–35 kHz. The end frequency for *E. bottae* is typically between 34 and 38 kHz and 50–53 kHz; *P. kuhlii* 38–41 kHz; and *T. teniotis* 14–16 kHz [21, 22].

All of the calls are within the range of previous studies in the Middle East [21–23]. **Table 2** lists frequencies for bats recorded from the Negev desert on the opposite side of Jordan. It clearly shows that calls reported in our study are similar for most species.

5. Bat caves in Jordan

A total of 142 caves and dwelling areas suitable for bats have been mapped (**Figure 5**). Caves are classified, as lava tunnel, pressure ridge, artificial, limestone,

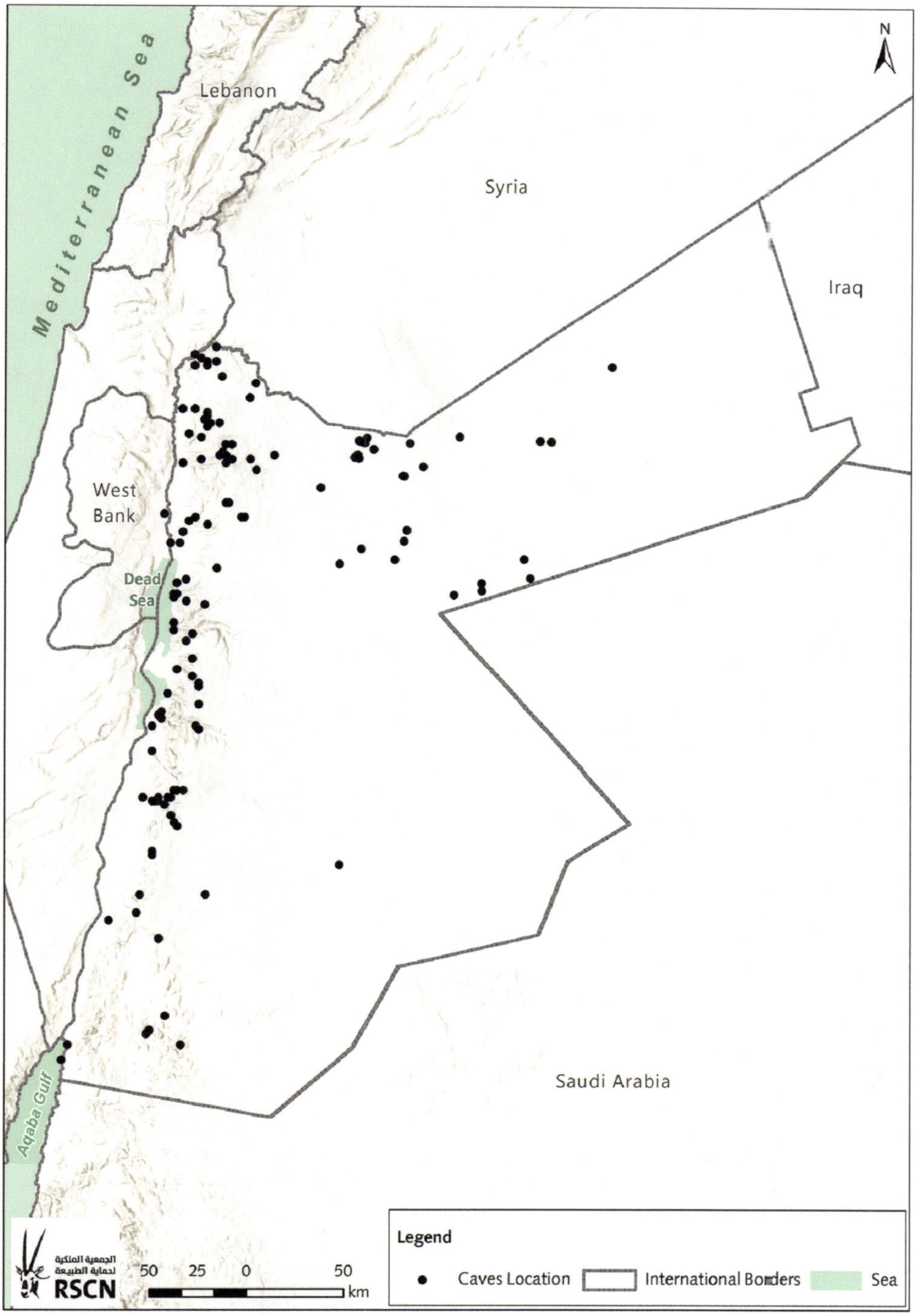

Figure 5.
Map of Jordan showing caves and other suitable bat-dwelling areas.

karstic, and sandstone caves. Most of the caves are located along the mountainous ridge extending from the north to the south on the western side of the country. Other sites include historical castles, mine-shafts, man-made tunnels, and rock crevices.

Based on previous studies over the past 30 years, the bat faunae for 28 caves are summarized in **Table 2**. The number of bat species per cave ranged from six to one, whereas Al Wardeh, Iraq Al Amir, Iraq Al Wahaj, and Zubiya caves harbored

Figure 6.
Some caves in northern Jordan. (A) Arjan cave entrance. (B) Arjan cave from inside. (C) Zubiya cave entrance after development. (D) Zubiya cave from inside. (E) Iraq Al Wahaj cave entrance. (F) Iraq Al Wahaj cave from inside.

six species. It seems that cave size is related to the number of bat species recorded; Zubiya, Al Wardeh, Dibbin Forest caves, and Iraq Al Wahaj are large caves extending over 300–500 m, while smaller caves usually are inhabited by a single bat species (**Table 2**). Other bat species such as *Barbastella leucomelas* and *Pipistrellus pipistrellus* were never observed to form colonies in caves in Jordan. The same is true for the Al Wardeh cave, which was entirely destroyed due to mining activities (**Figure 6**).

Our current projects on the bats of Jordan are to study the bat fauna associated with caves and identify threats that can affect bat populations. On the other hand, a campaign to protect bat important cave areas through legislative authorities is among high-priority issues.

6. IUCN conservation status of bats in Jordan

Eid et al. [24] compiled the national red list for the mammals of Jordan. Of the 26 bat's species, one, nine, and three species were listed as critically endangered,

endangered, and near threatened, respectively (**Table 3**). At the global level, *Rhinolophus euryale* and *Myotis capaccinii* are listed as near threatened and vulnerable, respectively.

Jordan is a member of several conventions, treaties, and agreements related to the provide protection for wildlife (Convention on Biological Diversity, Convention on International Trade in Endangered Species of Wild Fauna and Flora, Treaty of Conservation of Migratory Species of Wild Animals, Convention on the Protection of Marine Pollution by Preventing the Disposal of Waste and other Materials, RAMSAR, United Nations Convention to Combat Desertification). Several national laws have been enacted at the national level. In addition, the Agriculture Law No. 13 of 2015 addresses to a large extent Jordan obligations related to the protection of wildlife under these conventions [20].

Species	Common name	National IUCN status	Global IURCN status
Rhinolophus mehelyi	Mehely's horseshoe bat	CR	VU
Rhinolophus euryale	Mediterranean horseshoe bat	EN	NT
Rhinolophus ferrumequinum	Greater horseshoe bat	EN	LC
Rhinopoma microphyllum	Greater mouse-tailed bat	EN	LC
Taphozous perforatus	Egyptian tomb bat	EN	LC
Taphozous nudiventris	Naked-rumped tomb bat	EN	LC
Nycteris thebaica	Egyptian slit-faced bat	EN	LC
Miniopterus pallidus	Pale bent-wing bat	EN	LC
Myotis blythii	Lesser mouse-eared bat	EN	LC
Myotis emarginatus	Geoffroy's bat	EN	LC
Asellia tridens	Trident leaf-nosed bat	VU	LC
Rousettus aegyptiacus	Egyptian fruit bat	NT	LC
Rhinolophus hipposideros	Lesser horseshoe bat	NT	LC
Barbastella leucomelas	Asian barbastelle	NT	LC
Myotis capaccinii	Long-fingered bat	DD	VU
Rhinolophus blasii	Blasius's horseshoe bat	LC	LC
Rhinolophus clivosus	Geoffroy's horseshoe bat	LC	LC
Rhinopoma cystops	Egyptian mouse-tailed bat	LC	LC
Tadarida teniotis	European free-tailed bat	LC	LC
Eptesicus bottae	Botta's serotine	LC	LC
Hypsugo ariel	Desert pipistrelle	LC	DD
Myotis nattereri	Natterer's bat	LC	LC
Otonycteris hemprichi	Desert long-eared bat	LC	LC
Pipistrellus kuhli	Kuhl's pipistrelle	LC	LC
Pipistrellus pipistrellus	Common pipistrelle	LC	LC
Plecotus christii	Christie's big-eared bat	LC	DD

DD, data deficient; CR, critically endangered; EN, endangered; LC, least concern; NT, near threatened; VU, vulnerable.

Table 3.
Conservation status of bats in Jordan according to the global and national IUCN red lists.

Common name	Scientific name
Egyptian Fruit Bat	*Rousettus aegyptiacus*
Long-fingered Bat	*Myotis capaccini*
Natterer's Bat	*Myotis nattereri*

Table 4.
Bats listed in appendix III of bylaw no. 43 for the year 2008.

Regulation No. Z/2 for the year 2021 regulates wildlife protection, hunting, and trade. This bylaw was issued in accordance with article No. 56, of the Agriculture Law No (13) for the year 2015. Bylaw No. 43 for the year 2008 categorized mammals and other wildlife banned from hunting according to its level of protection (**Table 4**). This bylaw was issued in accordance with article No. 56, paragraph (H) of the Agriculture Law No (13) for the year 2015. In addition, Regulation No (Z/2) for the year 2021 includes instructions of regulating the International Trade in Endangered Species of Wild Fauna and Flora under Article (56) of the Agriculture Law No (13) for the year 2015 [20].

7. Threats affecting the bats of Jordan

7.1 Deforestation

Rhinolophus blasii, *Rh. euryale*, *Rh. ferrumequinum*, *M. capaccini*, *M. emarginatus*, and *M. nattereri* are inhabitants of natural forests in northern Jordan. Clearing forests for agricultural or housing projects is one of the major threats for bat populations, whereas bats loss roosting sites and feeding areas. This is true in Ajlun and Dibbin oak and pine forests since much of these forests are under development, where the noticeable decline was observed over the past decade, especially fo *Rh. ferrumequinum* in Dibbin Nature Reserve.

7.2 Urbanization

Over the past 70 years, the population of Jordan increased 17-folds during reaching 10,320,000 by 2021. A great burden on the natural and wild habitats due to this accelerated population increase became evident with the expansion of cities, towns, and villages. Construction style using cement was by far the most important factor that affected bat populations. Previously, old houses were built using mud, wood, or stone, creating suitable habitat for bat roosting. Many of these old houses are by now demolished and replaced by modern buildings [17]. Al Mahatta cave in Amman that used to harbor a population of the Egyptian Fruit Bat is by now void of bats due to housing projects that changed its integrity.

Noise and light pollution and heavy traffic disturbed many bats populations that are sensitive to human activity. One of the most evident examples of the disturbance that affected bat habitats is the disappearance of a healthy population of *Rh. ferrumequinum* was observed in 1959 in Swialeh [25], a township close to Amman. This area was entirely disturbed due to several forms of human activities including urbanization and population increase [17].

7.3 Tourism and vacationing

Tourism and outdoor activities in wild habitats (e.g., Dibbin Forests, Wadi Ram, and Zubiya) have a direct impact on the roosting populations of bat species living in

Figure 7.
Cave destruction in Jordan. (A) Asef cave remodeled as a recreation site. (B) Arjan cave with evidence of fire to be used as an animal barn. (C) Al Wardeh cave before mining activities in 2017. (D) Al Wardeh cave after mining activities in 2020.

such habitats. For example, large colonies of *Rh. ferrumequinum* that were common in Dibbin Forests, a national park visited frequently during vacations, disappeared due to outdoor activities in this park. By now, only a few individuals were observed in small caves and shafts. Hiking and cave exploration in Wadi Ram may contribute in declining populations of *E. bottae* and *Rh. hipposideros*.

The Egyptian Fruit Bat populations declined or disappeared from several sites where it used to be in abundance. Al Hemma cave is a classic example, where it used to harbor thousands of bats [7]. By now, only a few hundreds of the fruit bats were present and continued to decline. Also, the population of the Egyptian Fruit Bat in Wadi ben Hammad is declining due to extensive tourism activities within the vicinity of this site. The Zubiya cave, one of the largest karstic caves in Jordan, whereas seven bat species were previously recorded (*Rh. blasii*, *Rh. euryale*, *Rh. ferrumequinum*, *Rh. hipposideros*, *M. emarginatus*, and *M. nattereri*), was washed with high-pressure water and was closed by a gate preventing bats to gain entrance (**Figure 6C** and **D**).

Recently, many caves were turned into restaurants and coffee shops. This was observed in Asef cave where it was remodeled as a recreation site, and all bats were exterminated (**Figure 7A**).

7.4 Mining

In recent years, mining for extraction minerals for the cement industry and rocks for buildings expanded in many parts of the country. This in turn brought many bat populations to their demise. For example, Al Wardeh cave located near Ajlun was inhabited by large colonies of *Rh. blasii*, *M. blythii*, *M. emarginatus*, and *M. pallidus*

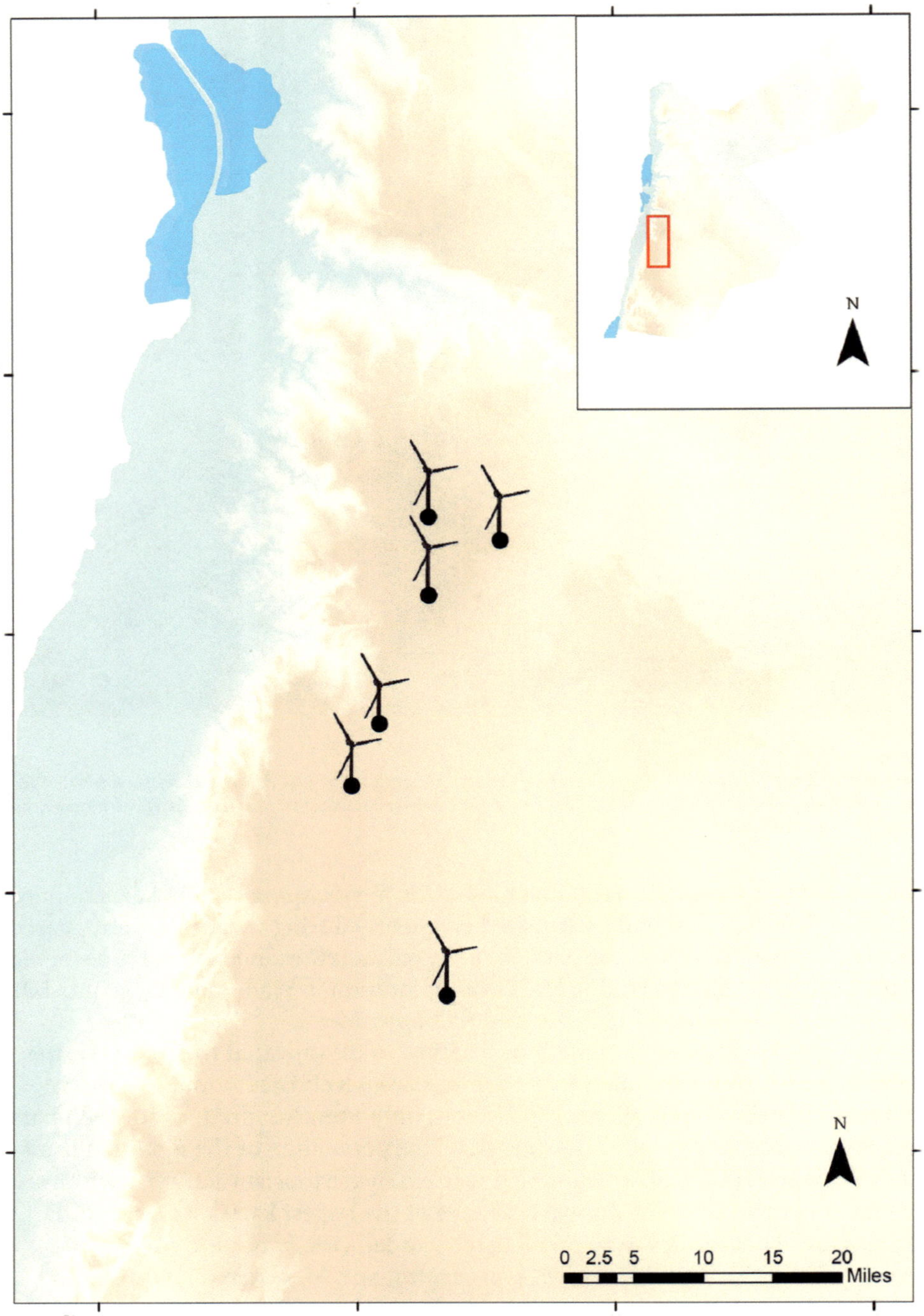

Figure 8.
Location of wind farms in southern Jordan.

for many years. We observed hundreds of bats in this cave all year round. Due to mining activity in this area in 2019, the cave is by now is destroyed and its main entrance has collapsed denying movement of bats in and out (**Figure 7C** and **D**).

Another cave used to harbor a significant population of over 500 individuals of *Rh. cystops* and *Rh. microphyllum* at Al-Majdal cave, located near Jarash, was destroyed and its entrance was closed.

Species	Species sensitivity	Eurobats level of collision risk	Likelihood of effect score	Risk rating
R. aegyptiacus	Low	Unknown	Low	Minor
Rh. blasii	Medium	Low	Negligible	Negligible
Rh. clivosus	High	Low	Negligible	Minor
A. tridens	Low	Unknown	Low	Minor
B. leucomelas	Low	Medium	Medium	Minor
H. ariel	Medium	High	High	Major
P. christii	Medium	Low	Negligible	Minor
N. thebaica	Low	Unknown	Low	Negligible

Source: IFC [26].

Table 5.
Species sensitivity, levels of collision risk, and risk rating for Wind Power Projects.

7.5 Wind power projects

In the past decade, windmills projects increased to reach up to six operational sites mostly in the southwestern part of the country, with a total of 151 turbines (**Figure 8**). Cumulative effect assessment was developed for Al Tafilah site to help in determining bat species that are at the highest risk and to identify potential mitigations and monitoring measures that should be considered by investors [26]. Eight species of bats were identified of major or moderate risk ratings (**Table 5**).

Carcasses of bats are surveyed on a monthly basis in the operational sites. We are aware of bat mortality in these sites; however, data remain undisclosed for the meantime.

7.6 Folk medicine and other practices

Bat blood is prescribed for female infants so no hair will grow under their armpits [27]. Fresh bat's blood is applied to the newborn body. This belief was adopted from the Roman culture and remained practiced by the local people in many parts of the Middle East that were under the Roman Empire, but is not practiced on a large scale.

Search for red mercury became a threat to bats in Jordan in the past 10 years. Many locals strongly believed that red mercury can be found in bat's nests. Caves and colonies in many parts of the country were disturbed in pursue of the red mercury to sell it at a high price. A campaign to educate the public that bats do not form nests through all forms of public media was undertaken.

7.7 Bats caves as animal barns

Scattered caves in the country have been used as sheep or goat's barns. Caves were sat on fire first to remove all kinds of wild animals such as snakes, scorpions, birds, and bats. This practice was observed in the caves of Jarash and Ajlun mountains (**Figure 7B**).

8. Bats and diseases in Jordan

Since the emergence of the Middle East Respiratory Syndrome Coronavirus (MERS-CoV) in Jordan in 2012 [28], health authorities and research institutes collected blood samples from the Egyptian fruit bat, *R. aegyptiacus*, and the dromedary camels. So far, only antibodies were detected in camels in Jordan, with no active virus, with a seroprevalence rate of MERS-CoV of 81% [29]. Swaps from bats were negative and no antibodies were detected. In Lebanon, HKU9-like viruses were detected in *R. aegyptiacus*, and serum samples tested from 814 bats were negative for MERS-CoV antibodies [30].

Although human rabies is considered very rare in Jordan, six isolates of rabies were found in wild animals (badger and squirrel) and domestic animals (cow, dog, donkey, and goat) [31]. No information is available on rabies caused by bats in Jordan. Further studies should evaluate the role of zoonotic infections that could be possibly transmitted by bats.

9. Conclusion

This study shed the light on the urgent need to conserve the bats of Jordan, taking into consideration the alarming decline in bat populations observed over the past decades affecting their natural habitats. Further studies on the ecological requirements and habitat selection for the bats of Jordan are needed. Such studies will provide baseline data to implement conservation strategies for each species. Other avenues of future research include identification, both quantitatively and qualitatively, the impacts of various man-made changes and threats to the existing bat population.

Acknowledgements

We would like to thank Qamar Almimi (RSCN) for map preparations. Also we thank Thabit Al Share, Sameh Khatatbeh, and Mohammad Al Zoubi (The Royal Society for the Conservation of Nature) for their help in the field studies. This study was supported in part by the Deanship of Academic Research, Jordan University of Science & Technology.

DOI: http://dx.doi.org/10.5772/intechopen.100407

Author details

Zuhair S. Amr[1]*, Omar A. Abed[2] and Mohammad Abu Baker[3]

1 Department of Biology, Jordan University of Science and Technology, Irbid, Jordan

2 Royal Society for the Conservation of Nature, Amman, Jordan

3 Department of Biology, University of Jordan, Amman, Jordan

*Address all correspondence to: amrz@just.edu.jo

References

[1] Qumsiyeh MB. New records of bats from Jordan. Säugetierkundliche Mitteilungen. 1980;**28**:36-39

[2] Darweesh N, Al-Melhim WN, Disi AM, Amr ZS. First record of the naked bellied tomb bat, *Taphozous nudiventris* Cretzschmar 1830, from Jordan. Zoology in the Middle East. 1997;**15**:13-14. DOI: 10.1080/09397140.1997.10637732

[3] Al-Omari K, Abu Baker M, Amr Z. First record of the Egyptian slit-faced bat, *Nycteris thebaica* E. Geoffroy, 1818, from Jordan. Zoology in the Middle East. 2000;**21**:5-7. DOI: 10.1080/09397140.2000.10637826

[4] Atallah SI. Mammals of the Eastern Mediterranean region: Their ecology, systematics and zoogeographical relationships. Säugetierkundliche Mitteilungen. 1977;**25**:241-319

[5] Atallah SI. Mammals of the Eastern Mediterranean region: Their ecology, systematics and zoogeographical relationships. Säugetierkundliche Mitteilungen. 1978;**26**:1-50

[6] Nader IA, Kock D. Notes on some bats from the Near East (Mammalia: Chiroptera). Zeitschrift für Säugetierkunde. 1983;**48**:1-9

[7] Amr ZS, Woodbury SC, Disi AM. On a collection of mammals from Jordan. Dirasat: Series B. 1987;**14**:131-136

[8] Bates PJJ, Harrison DL. New records of small mammals from Jordan. Bonner Zoologische Beiträge. 1989;**40**:223-226

[9] Amr ZS, Disi AM. Jordanian Mammals Acquired by the Jordan University Natural History Museum. Amman: Publication of the University of Jordan; 1988. 32 pp

[10] Qumsiyeh M, Disi A, Amr Z. Systematics and distribution of the bats (Mammalia: Chiroptera) of Jordan. Dirasat. 1992;**19**:101-118

[11] Qumsiyeh MB, Amr ZS, Al-Oran RM. Further records of bats from Jordan and a synopsis. Turkish Journal of Zoology. 1998;**22**:277-284

[12] Benda P, Sádlová J. New records of small mammals (Insectivora, Chiroptera, Rodentia, Hyracoidea) from Jordan. Časopis Národního Muzea Řada Přírodovědná. 1999;**168**:25-36

[13] Amr ZS, Qumsiyeh MB. Records of bat flies from Jordan, Libya and Algeria. Entomological News. 1993;**104**:43-46

[14] Qumsiyeh M, Schlitter D, Disi A. New records and karyotypes of mammals from Jordan. Zeitschrift für Säugetierkunde. 1986;**51**:139-146

[15] Karatas A, Weydan M, Sozen M, Shehab A, Amr Z. Karyotypes of bats (Chiroptera: Rhinolophidae, Vespertilionidae) from Jordan. Arab Gulf Journal of Scientific Research. 2008;**26**:193-198

[16] Abed OA, Abu Baker MA, Amr ZS. Seasonal habitat selection and temporal activity patterns determine the structure of Mediterranean bat assemblages. Zoology in the Middle East. 2019;**65**:11-19. DOI: 10.1080/09397140.2018.1540151

[17] Amr Z, Abu Baker M, Qumsiyeh M. Bat diversity and conservation in Jordan. Turkish Journal of Zoology. 2006;**30**:235-244

[18] Benda P, Lučan RK, Obuch J, Reiter A, Andreas M, Bačkor P, et al. Bats (Mammalia: Chiroptera) of the Eastern Mediterranean and Middle East. Part 8. Bats of Jordan: Fauna, ecology, echolocation, ectoparasites. Acta Societas Zoologicae Bohemicae. 2010;**74**:185-353

[19] Al-Eisawi D. Vegetation in Jordan. In: Haotot A, editor. Studies in the History and Archeology of Jordan. Vol. 1. Amman: Department of Antiquities; 1985. pp. 45-57

[20] Amr ZS. The Mammals of Jordan. 2nd ed. Amman: Al Rai Press; 2012. p. 308

[21] Hackett TD, Korine C, Holderied MW. The importance of Acacia trees for insectivorous bats and arthropods in the Arava Desert. PLoS One. 2013;**8**(2):e52999. DOI: 10.1371/journal.pone.0052999

[22] Hackett TD, Holderied MW, Korine C. Echolocation call description of 15 species of Middle-Eastern desert dwelling insectivorous bats. Bioacoustics. 2017;**26**:217-235. DOI: 10.1080/09524622.2016.1247386

[23] Benda P, Dietz C, Andreas M, Hotový J, Lučan RK, Maltby A, et al. Bats (Mammalia: Chiroptera) of the Eastern Mediterranean and Middle East. Part 6. Bats of Sinai (Egypt) with some taxonomic, ecological and echolocation data on that fauna. Acta Societas Zoologicae Bohemicae. 2008;**72**:1-103

[24] Eid E, Abu Baker M, Amr Z. National Red Data Book of Mammals in Jordan. Amman, Jordan: IUCN Regional Office for West Asia, Amman; 2020. p. 130. DOI: 10.2305/IUCN.CH.2020.12.en

[25] Harrison DL. Footsteps in the Sand. London: Ernest Benn Limited; 1959. p. 256

[26] International Finance Corporation. Tafila Region Wind Power Projects Cumulative Effects Assessment. Washington, DC, USA; 2017

[27] Aldebsawi AA, Al-Najdawi MM, Jarrar QB, Saleh MM, Moshawih SF. Middle Easterns' knowledge about bat blood use and effectiveness in preventing hair growth: A questionnaire survey. Archives of Pharmacy Practice. 2021;**12**:29-33. DOI: 10.51847/XdT8clmkcV

[28] Al-Abdallat MM, Payne DC, Alqasrawi S, Rha B, Tohme RA, Abedi GR, et al. Hospital-associated outbreak of Middle East respiratory syndrome coronavirus: A serologic, epidemiologic, and clinical description. Clinical Infectious Diseases. 2014;**59**:1225-1233. DOI: 10.1093/cid/ciu359

[29] Kandeil A, Gomaa M, Nageh A, Shehata MM, Kayed AE, Sabir JMS, et al. Middle East Respiratory Syndrome Coronavirus (MERS-CoV) in dromedary camels in Africa and Middle East. Viruses. 2019;**11**:717. DOI: 10.3390/v11080717

[30] Shehata MM, Chu DKW, Gomaa MR, AbiSaid M, El Shesheny R, Kandeil A, et al. Surveillance for Coronaviruses in Bats, Lebanon and Egypt, 2013-2015. Emerging Infectious Diseases. 2016;**21**:148-150

[31] David D, Hughes GJ, Yakobson BA, Davidson I, Un H, Aylan O, et al. Identification of novel canine rabies virus clades in the Middle East and North Africa. Journal of General Virology. 2007;**88**:967-980. DOI: 10.1099/vir.0.82352-0

Chapter 6

Dispersal Patterns, Mating Strategy and Genetic Diversity in the Short Nosed Fruit Bat *Cynopterus sphinx* (Chiroptera: Pteropodidae) in Southern India

Thangavel Karuppudurai and Steffi Christiane Ramesh

Abstract

The short-nosed fruit bat, *Cynopterus sphinx* is a common plant-visiting bat that is widely distributed throughout the Indo-Malayan region. In this chapter, we discuss the dispersal patterns, mating strategy and genetic diversity in the short-nosed fruit bat *C. sphinx*. We used a broad-range of techniques, including mark-recapture, radio-telemetry and molecular biology analyses. Our studies uncovered unique aspects of the dispersal, mating system and genetic diversity of these bats. Both the sexes of *C. sphinx* were found to disperse completely from the natal harems before subadult stage and young female *C. sphinx* become members of a harem much earlier than their male counterparts. The nonharem males are reproductively active, gain access to harem females and sire more offspring in July–August breeding season than March–April breeding season and presumably obtain some reproductive success. Our molecular study shows that considerable genetic diversity was observed in this species from different zonal populations, possibly due to complete dispersal of juveniles of both the sexes from their natal groups and gene flow between the zones. All these studies suggest not only a predictive framework for future studies, but also the use of these data in the management and meaningful conservation of this species.

Keywords: *Cynopterus sphinx*, fruit bat, dispersal pattern, mating strategy, genetic diversity

1. Introduction - Study species

The Indian short-nosed fruit bat, *Cynopterus sphinx,* belongs to the Old-World fruit bats (Megachiroptera: Pteropodidae) (**Figure 1a**). It is a common plant-visiting bat that occurs throughout the Indo-Malayan region and roosts solitarily or in small groups in the foliage [1]. It weighs about 45–70 g and lives in small clusters of about 3–30 individuals [2–4]. Dog-shaped head, divergent nostril, large prominent eyes and short ears with white margin are the unique morphological characters by which one can easily identify this species. In males, the chin, anterior part of shoulders, sides of the chest, belly and thighs are characteristically orange tinted

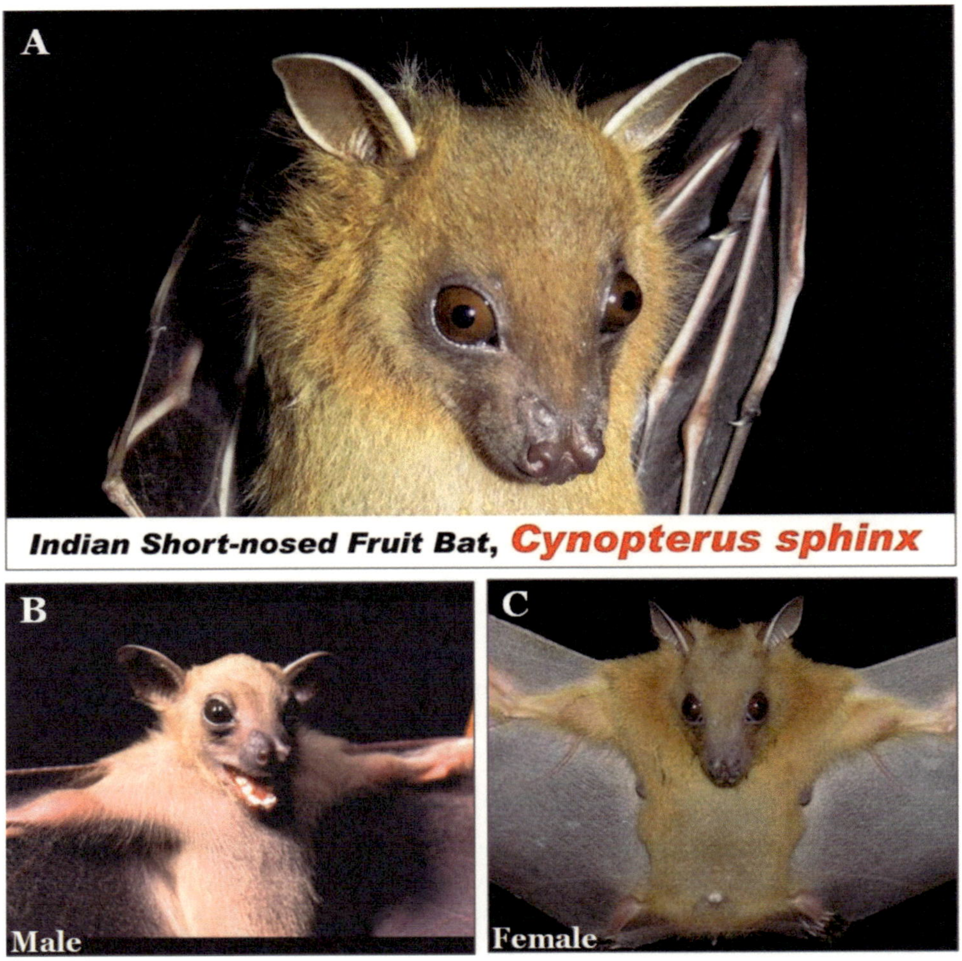

Figure 1.
The study species Indian short-nosed fruit bat (a) Cynopterus sphinx. A close view of (b) male and (c) female.

(**Figure 1b**). In females, the collar is usually tawny brown, the rump is gray brown and the belly is paler gray with slightly lighter hair tips (**Figure 1c**).

These bats use several types of diurnal roosts and are known to alter different types of foliages (palm and mast trees) to construct tents and attract females (**Figure 2a** and **b**) [2, 4–6]. Although different types of altered plant structures are referred to as bat tents [7], the first account of tent making by a male bat came from observations on *C. sphinx* [2]. During the breeding season, most of these bats live in groups called "harems" (**Figure 2c** and **d**) consisting of a single male and more than one female [2–4, 6, 8]. Harem males defend such tents against other males and thereby enabling copulation with a large number of females which is the primary mating strategy adopted by *C. sphinx*. They follow polygynous mating system (prolonged association of one male with more than one female) based on resource, called resource defense polygyny. The recruited females are defended as harem by a single male [2, 6]. However, apart from such successful males, a number of adult males were also observed roosting solitarily [8–10].

C. sphinx is a polygynous-mating bat that has a polyestrous reproductive cycle with two well-defined and highly synchronous parturition periods per year [11, 12]. Typically, in India, parturition takes place during the month of March/April and July/August. Females can reproduce a maximum of two pups in a year [11, 12]. Females endure a postpartum oestrus once the young ones are born during the month of

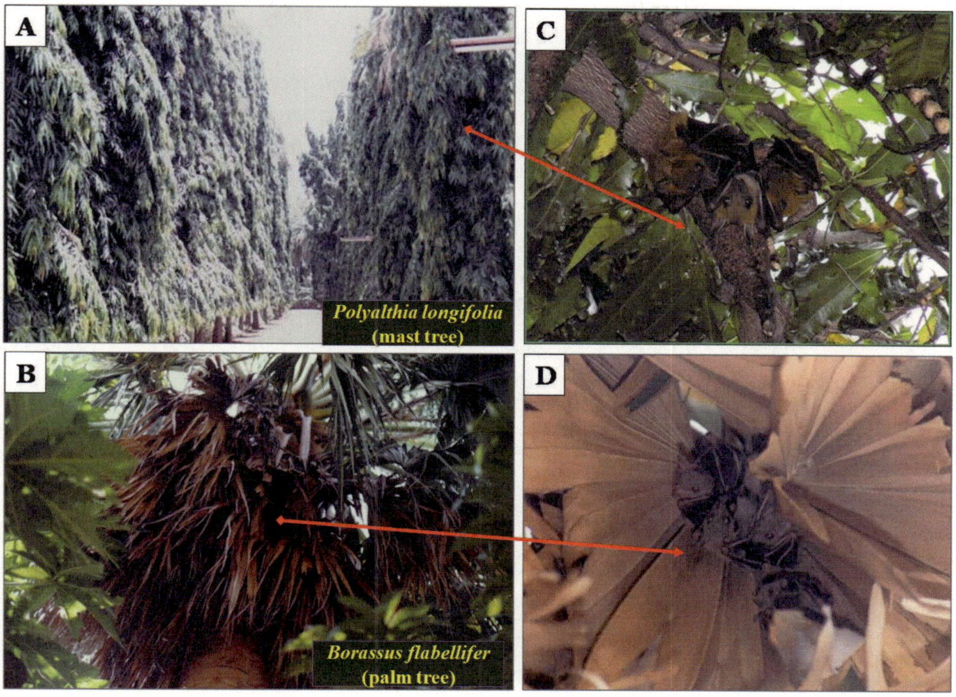

Figure 2.
The major roosting trees (a) Polyalthia longifolia (mast tree), (b) Borassus flabellifer (palm tree) commonly used by the Indian short-nosed fruit bat C. sphinx in South India. A closer view of group of C. sphinx roosting in tents of (c) mast and (d) palm tree. Arrows indicate the bat roosts.

March/April [11, 13]. To wean the cohorts born during March/April, females lactate the young ones while they are pregnant. Females remain anoestrus until October after the birth of July/August cohort [14, 15]. Neonates weigh *ca.* 11 g at birth and begin to fly at approximately 40–50 days of age, although young may continue to suckle from their mothers for 10–15 days after they initiate flight [12, 16]. At weaning, young *C. sphinx* weight about 51% of adult body mass and achieve adult body dimensions at approximately two months of age [1]. In this chapter, we discuss dispersal patterns, mating strategy and genetic diversity in the short-nosed fruit bat *C. sphinx* (Chiroptera: Pteropodidae) in southern India.

2. Dispersal patterns

2.1 Introduction

Displacement of a juvenile from its birth place to the first site of reproduction is termed as natal dispersal [17]. This natal dispersal is one of the factors contributing to the central evolutionary forces that affect the natural populations. Also, it is the key life history trait that is involved in both species persistence and evolution [18]. Understanding the patterns of dispersal is important in population ecology and conservation biology [17, 19]. Bats are known to exhibit varying degrees of dispersal and philopatry based on their social system [20, 21]. The short-nosed fruit bat *C. sphinx* is a group living, frugivorous, megachiropteran bat. In *C. sphinx*, the juveniles of both the sexes are believed to disperse completely from the natal harem [3]. However, their status after dispersion remained unanswered. Do they join established harems immediately or aggregate to form a new harem? Answering this question is the aim of this section.

2.2 Materials and methods

In order to understand the dispersal pattern in *C. sphinx*, the mark-recapture study was carried out in the day-roosting places in and around Palayamkottai, Tamil Nadu, South India (lat: 8° 44′ S; long: 77° 42′ E). We spotted day roosts by citing the accumulation of leaf pellets, rejected fruits, seeds, and leaves under trees especially *Borassus flabellifer, Polyalthia longifolia Washigtonia filifera, Caryota urens,* and *Vernonia scandens.* When any of these accumulations are seen, it is very likely that *C. sphinx* bats roost. Randomly 4–6 day roosts were censused every week. Bats were captured just before emergence with the help of a hoop net attached to an extendable aluminum pole. The entire tree was enveloped with a 6 m x 9 m nylon mist net (Avinet-Dryden, New York, USA) to prevent bats from escaping. The mouth of the hoop net was placed at the entrance of day roosts. A minor disturbance was caused to the trap the bats inside the tent. For each bat, sex, age, forearm length and body mass were recorded [16]. After processing, the bats were held in net cages and were released at their roosts on the evening of the same day they were captured. Only those colonies, which were captured completely, were included in the present study. The proportion of males and females in different age classes viz. pups, juveniles and subadults were estimated and their roosting patterns were recorded.

2.3 Results and discussion

Our results showed that female proportion in harem increases considerably from pups to subadults. Both the sexes are equal in number (1:1) in the pup stage, whereas the sex ratio was female biased in the juvenile (1:1.8) and subadult stage (1:4.2). In mammals, dispersal is usually male biased and this also holds true for most bat species studied to date [17, 22, 23]. However, several studies on tropical species indicate that there may be cases where females also disperse [24]. Our study suggests that the juveniles of both the sexes disperse from their natal group before entering the subadult stage. We mostly captured dispersed juveniles and harems in which post-lactating females were present without the young ones. In many of the day roosts, the number of juvenile bats were disproportionate to the number of post-lactating females, especially when the juvenile bats were predominant.

The capture rate of juvenile females outnumbered the juvenile males, which suggest that the males dispersed early from the natal roost. One could suggest that maternal neglect could be playing a strong role in the altered sex ratio in *C. sphinx*. If mortality is a factor at the pre-weaning stage, both the sexes are expected to suffer equally: whereas if adults eject the males by force, then there may not be any juvenile males in their parental roosts. Contrarily we observed some juvenile males roosting in the natal roost though the number was much less when compared to females. Similar studies in other bats show that young females of short-tailed fruit bat *Carollia perspiscillata* are more likely to disperse away from the natal roosts than the young males [25]. In *Nycticeius humeralis,* all juvenile male bats deserted the roost faster than females and they were neither seen again in the roost nor foraging nearby. Contrary to this, juvenile females remained in the foraging area with their mothers after recruitment. They evidently continued to nurse for about three weeks longer, since milk could be expressed from the mammary glands of captured adult females until mid-July [26]. In a Neotropical bat *Lophostoma silvicolum* both male and female offsprings disperse before maturity and the polygynous mating system may lead to all-offspring dispersal more often than previously assumed in mammals [27].

In accordance with the general mammalian pattern the females of most group-living bats, including some harem-forming species and all temperate zone species, are strongly philopatric which includes evening bat *N. humeralis* [17, 26, 28], vampire bat *Desmodus rotundus* [29], the brown long-eared bat *Plecotus auritus* [30, 31], Bechstein's bat *Myotis bechsteinii* [32–34], greater horseshoe bat *Rhinolophus ferrumequinum* [35], mouse-eared bat *Myotis myotis* [22] and northern long-eared bat *Myotis septentrionalis* [36]. A typical female philopatry was not observed in the group living *C. sphinx*. In this study, juveniles of both the sexes were found to disperse completely from their natal harems. Several reasons for female dispersal have been postulated [37]. It is often assumed that dispersal costs are higher for females than for males, whereas the benefits are thought to be higher for males. Moreover, absence of female philopatry is not uncommon among harem forming Neotropical bats [25, 29].

In this study, the female proportion in harems increases considerably from pups to subadults. Compared to juveniles (1:1.8), sex ratio was highly skewed towards females in the subadult stage (1:4.2). From the total of 52 subadults, only 10 were males. Moreover, not a single subadult male bat was found to be roosting in a harem. This explains that the males disperse from the natal roost before subadult stage. From the 42 captured subadult females, it was observed that four subadult females were pregnant showing that the females matured earlier and were engaged in reproductive activities during the subadult stage itself. Lower rate of juvenile survivorship maybe one of the reasons for the low capture of adult males [12], which remains unclear. The probability of censusing these bats remain low because the male bats preferred to roost in dense, unmodified and previously unoccupied foliages. During the capture, the number of subadult females were larger as they joined established harems, formed a new harem of subadult females with an adult male or they remained alone in the roost.

However, we were not able to find out whether the dispersed juvenile bats return back to their natal harems. The probability of making local migrations even during the breeding season is rare in both *Rhinopoma hardwickei* and *P. auritus* and hence they show greater fidelity towards their roost sites when compared to other bat species [38, 39]. However, the degrees of roost fidelity differ according to the sex and age in different bat species. During the nursing season female *Miniopterus schreibersii* bats strongly exhibit philopatric behavior since they returned to their place of birth to produce young ones [40]. Also, it was observed that the males extended greater level of connection to their birth site and also the juvenile female bats showed higher fidelity to their natal roost than did males and bats of other ages [41]. Moreover a particular population of *C. sphinx* from Pune, there were no recruitment of females to natal harems and also these females did not join other harems in the same or neighboring colonies [3]. Young females joining harems have been documented in bats like *Artibeus jamaicensis* and *Phyllostomus hastatus*. Though, technically the mating system of both these species appears to be a form of polygyny similar to *C. sphinx,* all three species seem to differ in the way young females are recruited in to a harem. In *A. jamaicensis*, it was observed that harems contained females of all ages, which suggests that subadult females join established harems [21], whereas, harem of the greater spear-nosed bat *P. hastatus* forms a new generation of subadult females [20]. However, in *C. sphinx* the harems consist of all age groups of harem females as well as the harems with only subadult females were in common.

2.4 Conclusion

We identify that young female *C. sphinx* become member of a harem group much earlier than the male. The reason for nonharem young females staying near to the established harems is not known. But this roosting pattern may improve the chances of breeding attempts made by young females as they attain sexual maturity earlier (*ca.* 7 months) compared to males (*ca*. 18 months) [12, 42].

3. Mating strategy

Factors responsible for the occurrence of nonharem males and mechanisms used to acquire harem male status.

3.1 Introduction

Polygynous mating is one of the most salient features of mammalian social structure and has potentially far-reaching consequences for a diverse array of evolutionary processes [43]. Male reproductive success in polygynous mammals is largely attributed to the spatial and temporal patterns of female aggregation [43, 44]. Receptive females are reliant on variation in resource distribution, predation pressure, costs of social living and activities of males [43]. One of the major factors that affect the mating success of resource-based polygynous mating animals is the resource distribution. Females choose males indirectly by mating with males that defend the highest quality resource when the males control access to the resources that these females require for reproduction [44]. Males that can make the greatest genetic contribution to the fitness of their offspring are chosen by the females [45].

Bats exhibit various forms of mating behavior ranging from simple monogamy to resource and female defense polygyny, as well as leks [46, 47]. Among these, resource defense polygyny is the most commonly observed mating pattern [48]. Bats establish a harem by defending critical resources such as food, shelter or mates [46]. Males potentially gaining favored access to several females is considered as one of the main benefits in resource defense polygyny and a healthy male inseminates the females [43, 46]. It is observed that several Neotropical fruit bats such as *Uroderma bilobatum* [49, 50], *Vampyressa nymphaea* [51], *Ectophylla alba* [52], *A. jamaicensis* [53–55] and *C. sphinx* [2] follow this type of mating strategy. Most of the bat species often spend the day and a large portion of the night in the roost which shows their attachment to their roosts [7]. However, it was observed that out of 1,300 bat species, only about 20 species are known to make their own roosts [56]. A striking feature of some polygynous bat species is that they often alter different types of foliage to create tents [2, 3, 57].

C. sphinx is known to exhibit polygynous mating system (i.e. prolonged association of one male with more than one female) based on resource defense polygyny [4]. In *C. sphinx,* the adult males are categorized into two groups, harem and nonharem males. Males construct and defend tents (resource) and recruit females to gain mating access and this organization is called a harem. In order to attract females, the harem males defend critical resources during the breeding season and this type of system is called harem-polygynous mating. Though, several studies suggest that the breeding population of *C. sphinx* also consists of nonharem males which dwell in roosts adjacent to harems [3, 4].

If this is true, what is the role of such nonharem males in the population or colony? What are the factors that cause the occurrence of nonharem males in a colony of *C. sphinx*? Are trees and foliage suitable for tent-making a scarce resource? Are solitary males less competitive and so remain isolated from the breeding activities? However, the factors responsible for the occurrence of nonharem males and mechanisms used to acquire harem male status are not understood fully.

3.2 Materials and methods

The study involves understanding the mating status of solitary males, bat captures were confined to day roosting places. Captures at roosting places indicated whether a male was solitary or a harem holder. Every week, we inspected trees and censused day roosts regardless of the number of incumbents (solitary or harem). Roosting groups with a single adult male with one or more adult females were considered as harems [46]. However, apart from such successful males, a number of adult males were also observed roosting solitarily. All the individuals of harems and the solitary males, which roost adjacent to the harems, were captured just before emergence using a hoop net with an extensible aluminum pole.

All the bats of harem groups and nonharem males were tagged with a color-coded bead necklace. We used beads of ten different colors, each color denoting a number from 0 to 9. We loaded each necklace with 1–3 beads. Thus, there were 999 possible sequential arrangements of the color beads. We have used this type of tagging for various studies and have observed no apparent detrimental effects on bats. After marking, all individuals were released at the site of capture. These color coded bead necklace markings allowed us to identify individuals and determine their previous roosting locations. The census, mark-recapture and radio-telemetry studies data were used to assess the reproductive condition, mobility, roosting pattern and status of adult males (harem/nonharem).

3.3 Results and discussion

One of the striking features of tent-making male bats is that they use tents as a resource to recruit large numbers of females and copulate with them [46, 51]. Although experimental evidence supporting causal factors for resource-defense polygyny is lacking, scarcity of resources is thought to be one of the factors for aggregation of females [58]. Solitary roosting existence of some adult males is one of the main consequences of resource-defense polygyny as the males fail to defend a resource. We attempted to study the resource-defense polygyny in *C. sphinx*. We observed that there were no shortage of roosting sites and these solitary males remained reproductively active. Similar, results have been reported in a Jamaican fruit-eating bat *A. jamaicensis* [59, 60].

Our results suggest that the male success in female recruitment was not due to shortage of tents. We found that, nearly 39% of adult males were roosting alone. This observation was based on >90% of nonharem males roosting adjacent to harems and also 50% of nonharem males had scrotal testes. In addition, the mark recapture study showed that the transition status of males from nonharem to harem was possibly due to previously unobserved mode and the female recruitment is associated with resource (roost). It indicates that the solitary males are involved actively in female recruitment to their roosts and also in the process of mating. During our study we observed that many solitary males recruited females within a short period of time. A nonharem male's effort to gain access to adult female cannot be hindered by the solitary nature *per se*, which is found reliable based on expectation. Since it is a choice of every animal to represent itself genetically superior among a population, we believe that the breeding behavior of the animal is not restricted by the resource.

The mode of attaining harem male status differs from species to species. For e.g. young males of *P. hastatus* are known to gain access with harem females if the harem male dies or gets displaced [61]. It was observed that under captive conditions, two adult *C. sphinx* males, competing with each other to take control of a tent resulted in the harem male turnover. However, the turnover of harem male did not have any impact on the cohesion of harem females [62]. In other harem forming bats, males exhibit a typical pattern during their ascendance to the dominant status. A size-based hierarchy for males in the social system was observed in *A. jamaicensis* [59, 63], with some larger harems being occupied by a small sized subordinate male apart from a dominant male. In *S. bilineata*, some individuals are associated in harems over several years and the non-territorial males build up site-specific dominant hierarchies [64] and for coalitions of male manikins (*Chiroxiphia linearis*). In order to attract females, the subordinate males perform costly displays by playing a satellite role. But they do not obtain reward immediately because the dominant male practically fathers all the young bats [65].

However, subordinate manakin males readily take the place of the dominant males in order to obtain a long term benefit from the association. Subordinate males' relationship with dominant males adds an extra benefit to the subordinate males by increasing the inclusive fitness thereby leading to higher reproductive output [66]. Our efforts to identify the morphological differences between harem males and nonharem males were not successful as we did not find substantial differences in the forearm length and body mass. This is surprising because an individual's body condition is often the most important determinant for alternative mating tactics [67, 68]. Larger and heavier males are typically dominant in male–male contests and reproduce more often [69]. In male common shrews *Sorex araneus,* the differences in body weight may vary with age [70]. Body weight of the males that differed in mate-searching behavior varied during the early stage of maturation while no significant difference was observed in the body size of both the types of adult males. In the present study, it was observed that most of the adults with T3 tooth-wear class were harem males, while the ones with T1 tooth-wear were found to be in the group where nonharem males existed. Therefore, among the first breeders, a strong competition for mates takes place. Observation made it evident that approximately 50% of males with T2 tooth-wear class were harem males while the remainders were found to be nonharem males. Also, these results suggest that harem males and nonharem males differ in age slightly.

Individuals with territory and resource, typically have a higher reproductive success than the males without territories due to strong competition for mates in a polygynous mating system. Males with territories usually monopolize and probably fertilize many females [71]. The males which does not possess any territory follow alternative mating strategy either as satellites [72, 73] or as sneakers [74, 75]. Similarly, among polygynous bats such as *Parasenecio hastatus* [20], *D. rotundus* [76] and *Saccopteryx bilineata* [77] it has been observed that the harem males monopolized the females incompletely. Chances of nonharem males fertilizing the harem females increased due to the incomplete monopolization of harem females by harem males. However, the mode of nonharem males gaining access to harem females was not clearly known to determine the "alternative strategy" [58].

The roosting preference of females seems more likely to increase the chances for nonharem males fertilizing some of the females. Apart from the mating success of nonharem males, low paternity for harem males can also occur as a result of female choice. *S. bilineata* are highly mobile and actively select their roosting location. During the course of a day, some females are found to shift their roosting territories while others disperse to other colonies. Importance of female choice specifically in highly mobile animals with harem system was reported [77]. Our radio-telemetry studies suggest that 3-postpartum *C. sphinx* females were found to be visiting a nonharem male especially during the night hours and also were involved in mating. Females were periodically shifting their tents since fluctuations were observed in the harem size on a day-to-day basis [2]. Similarly, movement of females between harems has also been observed among the polygynous bats such as *A. jamaicensis* [60, 78], *P. hastatus* [20, 24], *D. rotundus* [76] and *S. bilineata* [77, 79].

Similarly, pallid bats *Antrozous pallidus* switch roosts without respect to group structure during pregnancy, but invest energy in communication to move as part of a cohesive group during lactation [80]. Storz et al. [4] reported that in *C. sphinx*, new harems are formed when parous females from an established harem join a previously solitary male in a different roost within the same colony. Reproductive success of male *C. sphinx* depends on colony structure during the previous post-partum estrus rather than current parturition [58]. Similar results have been reported in round-eared bat, *L. silvicolum* [47].

In the Bechstein's bat *M. bechsteinii* female bats frequently change between many different roosts [32]. In *M. myotis*, 16% of 435 ringed females appeared at least once in foreign colonies and about 6% switched colonies permanently [81]. Furthermore, in *P. auritus* [39] and in *R. ferrumequinum* [82] females occasionally switch colonies. Moreover, the permanence of roost sites like caves, mines and buildings, in contrast to the ephemeral nature of sites like trees, could account for patterns of roost switching observed in free-ranging bats [80]. It is well known that tree-roosting bats switch roost sites every few days, but the motivation underlying roost switching is not well understood [83]. The roost switching may reflect the maintenance of long-term social relationships between individuals of the larger colony [84]. We observed that the females of *C. sphinx* move between two or more roosts and also, many harems were completely abandoned the tents. In response to changes in access to diurnal roosts with suitable microclimates or the availability of fruit and nectar resources, female *C. sphinx* may alternate between different roosting habitats in the same local area [4], although they do not undergo seasonal migration. A similar pattern has been observed in a breeding population of the Neotropical fruit bat *C. perspicillata* and *Pteropus poliocephalus* [85].

C. sphinx has two well-defined parturition periods per year [11, 12, 86]. A huge difference was found between these two parturition periods while assessing the paternity of harem males [58]. The authors attributed this difference between the parturition periods to the availability of roosting sites associated with seasons. The survey also showed that average size of harem was found to be slightly higher during the dry season than during the wet season. However, we observed little correlation between harem size and availability of roost sites. In the study area, during both wet and dry seasons, bats abundantly roosted in both mast trees and palm trees. Shortage of roosting sites were rarely observed in our year long survey. Throughout the year, more than 90% of the day roosts were occupied by *C. sphinx*. Similarly, food resources in the area were also very vast to influence harem size [87].

During July and August, the frequency of nonharem males were found to be highest. Timing of sexual maturity of young males might be a probable reason, though no reports on timing of sexual maturity of young male bats in southern India. Reports from central India suggest that males born during the June–July parturition and February–March parturition were able to mate during September–October of the following year [88]. The number of nonharem males censused during August to October was relatively high in the study area. This can be attributed to the competition among first time breeding males to establish a day roost to recruit females before securing mating in October–November.

Our radio-telemetry observations suggest that females aggregated with a solitary male. Interestingly, aggregation occurred only after the male occupying a tent which was probably constructed by another male. Our tagging efforts might have probably disturbed the harem but the exciting aspect of this observation is the subsequent female aggregation and the way by which a solitary male succeeded in recruiting females. In a short span of time it may not be possible for a male to succeed in mating, if it followed the primary strategy involving construction and defense of tent leading to female recruitment. In addition, we observed that the solitary male spent less time away from the roost at night after female recruitment by frequently visiting the roost throughout the night and by making several short foraging flights spaced randomly throughout the night [10]. This behavior is consistent with the earlier reports on the activity of harem males in *C. sphinx*, *A. jamaicensis, P. hastatus*, *C. perspicillata* and *Balionycteris maculata* [2, 24, 25, 54, 89]. This suggests that in *C. sphinx,* some type of territoriality is associated with shelter [61]. Situations under natural conditions like displacement or death of harem male can also occur though, roost abandonment of harem appears to be artificial. Among the species, the mode of attaining harem male status differs. In *A. jamaicensis* [59], *S. bilineata* [64] and *P. hastatus* [61], a size based hierarchy was observed. However, in *C. sphinx,* no such pattern has been reported. In order to attract maximum number of females (extreme variation in the group size) and hold them together, some recognizable factors should be considered. But, morphological features of males [90] and the characteristics of tents [2], does not influence in female recruitment.

3.4 Conclusion

Although the high clustering of females in confined roosting places appears to facilitate resource-defense polygyny in *C. sphinx,* recent results showed the failure of harem males in the exclusive defense of harem females and the morphological variables did not differ between harem and solitary males. The present observation suggests that the female recruitment is associated with resource (roost). Taken together with the present results of reproductively active nonharem males, it seems that the solitary nature of some adult males in the population of *C. sphinx* may not be a forced option. However, further investigation is necessary to find whether the solitary adult males adapt any alternative reproductive strategy to usurp mating opportunities of harem males. In order to understand the complex mating strategy of *C. sphinx,* extended molecular genetics techniques to behavioral ecology is required.

4. Molecular genetic analysis of mating strategy

4.1 Introduction

As our understanding on mating systems increases, it becomes obvious that apparently species-specific mating behaviors often vary both between and within population [91]. Reproductive strategies are shaped by natural selection favoring individual with the greatest lifetime reproductive success. However, not all mature individuals adapt to the same reproductive strategies [69]. When competition for access to mates is severe, young reproductive individuals sometimes opt for alternative mating behaviors. Environmental or demographic factors may constrain the number of males that were able to employ the most successful strategy [92]. Alternative tactics in reproductive behavior enable individuals to maximize their fitness in relation to competitors of the same population. Among polygynous mammals, territorial behavior is almost exclusively a male trait believed to function primarily as a reproductive strategy to secure mates. Because mammals are committed to their progeny through gestation and lactation, female reproductive success usually is more readily quantified than male reproductive success. Male reproductive success in polygynous mammals is largely attributed to the spatial and temporal patterns of female aggregation [43, 44, 91].

Most known mating associations in bats are composed of a single male and several females and such organization are usually called harems [46]. *C. sphinx* is known to exhibit polygynous mating system (that is, prolonged association of one male with more than one female) based on resource availability and such behavior is popularly known as resource defense polygyny [4]. Though, several studies have shown that the nonharem males also occupy the roots nearby harems most of the time [4, 8–10, 93]. Although, the role of nonharem males as probable fathers has not been studied well in *C. sphinx* population.

4.2 Materials and methods

Bats were collected from the foliage tents of *P. Longifolia* (mast tree) and *B. labellifer* (palm tree) using a hoop net with an extensible aluminum pole. Bats were sampled over a period of four weeks immediately following each of four annual parturition periods: March–April (dry season) and July–August (wet season). A medical punch will be used for the excision of tissue (4 mm^2) and care will be taken to place it in an area between the blood vessels to avoid injury (wing membranes healed within 3–4 weeks [93, 94]. After each sampling, the punched hole and the punch will be disinfected with 70% ethanol. No negative effects of this treatment on the health of the bats will be observed. It should also be noted that the bats frequently have natural injuries of this type in their wing membranes.

The collected blood samples will be immediately mixed with Anticoagulant Citrate Dextrose (ACD), transferred to microcentrifuge tubes and sealed with parafilm. The blood and tissue samples will be stored in ice, transported to the lab and stored at −20°C until DNA extraction [93, 94]. No bats will be killed or retained as specimens during this project. We will be following the Institutional Ethical and Bio-safety Committee Guidelines of Madurai Kamaraj University. PCR based RAPD strategy was used to study the paternity of harem males and nearby nonharem males to the young born in the harems.

4.3 Results and discussion

During the wet (July–August) season, we captured 27 harem males, 30 nonharem males and 125 offsprings were analyzed to assign the reproductive success of harem and nonharem males. Out of the 125 offsprings the nonharem males sired 73 offsprings (average 58%) and the harem males sired only 52 offsprings (average 42%). During the dry (March–April) season 14 harem males, 18 nonharem males and 142 offsprings were captured and analyzed to assign the reproductive success of harem and nonharem males. Of the 142 offsprings the harem males sired 132 offsprings (average 94%) and the nonharem males sired only 10 offsprings (average 6%). From these results, we identified that the reproductive distribution is unequal between harem and nonharem males. It indicates that the harem males failed to control harem females thereby increasing the chances of nonharem males to fertilize some of the harem females. In addition, in southern India, during the dry season the spatial dispersion of female *C. sphinx* is highly clumped due to limited roosting sites and the harem male sires 96% of offspring conceived during this period [58]. In total contrast during the wet season, more roost sites are available and females are dispersed more widely. In this case, the harem male sired only 40% of offspring, while the other 60% offsprings were sired by other (solitary) males. The possible movement of females between harems was suggested as one of the reasons for this observation. Similarly, among the polygynous bats *A. jamaicensis* [59, 78], *P. hastatus* [20], *D. rotundus* [76] and *S. bilineata* [77, 95], incomplete monopolization of females by harem males has been observed. The harem males failed to control the harem females as result the increases the chances for nonharem males to fertilize some of the females.

The most commonly described mating system in bat species is polygyny, in which males defend a resource to recruit and have exclusive mating access with a large number of females. The resource may be a foraging area or a roosting site or the females themselves. However, several genetic analyses have shown that paternity is biased in polygynous mating systems. For e.g. a paternity study in *S. bilineata* demonstrated that 71% of offspring born into a harem are not sired by the resident harem male, but are instead fathered by non-territorial males [77, 95]. Similarly, in *P. hastatus*, harem male fathered 60–90% offspring [20], while the harem male in *D. rotundus* fathers approximately 45% of young [76] and the estimated paternity for dominant males of *A. jamaicensis* ranged from 33 to 83% [78].

4.4 Conclusion

The molecular genetic analysis of mating strategy assignments based on RAPD results suggest that during July–August breeding season (wet), the nonharem males gained access to females and sired more offspring than March–April breeding season (dry). These results suggest that nonharem males are reproductively active, gain access to harem females and enjoy some reproductive success. To understand the reproduction of nonharem males, further investigations are necessary. Solitary behavior can be an acceptable alternative to territoriality because the reproductive success of some nonharem females were relatively high. Solitary males sired number of juveniles but had no costs for roost defense. Harem males were not able to control the movement of the females in their harems because reproduction by nonharem males is possible [2–4, 77, 93]. Since, harem females provided no parental care, the females were allowed to choose their mating partners. The behavior and reproductive success of nonharem males over their lifetime could clarify whether they potentially compensate lower reproductive success per year with longer persistence in the harem.

5. Genetic diversity within and among populations of *C. sphinx*

5.1 Introduction

Genetic variation is an important factor in determining the ability of a species to adapt to new environmental conditions and therefore may be an important measure of the evolutionary potential and long-term viability of a species. The information on the amount of genetic variation within a species and its distribution within and between populations would aid in bat conservation planning [96, 97]. To understand both the past and current behavioral processes, it is vital to know the population structure of a species. Colonization and/or dispersal events can be inferred by characterization of population structure at the macro-geographical level, while social organization within a population can be used to infer the micro-geographical structure [30]. Both direct (mark-recapture studies) and indirect (genetic) techniques [98] should be used to study the population structure of individuals to understand the degree of spatial variation both in distribution and genetic composition [99].

In general, the high dispersal abilities are associated with a low population structure [100], which has been reported for some mobile species, including birds [101] and bats [102]. Studies describing molecular patterns of intraspecific geographical differentiation in bats have indicated a low level of genetic divergence and a limited geographical structure in species with continental distribution [103]. However, high-intraspecific divergence levels with clearly defined geographical structuring have also been observed. These different results can be attributed to the different molecular markers used in the various studies. Studies on different bat species using the molecular genetics approach have shown genetic diversity among distant populations [102–105].

In *C. sphinx*, the behaviors of tent construction [5], reproduction [11], foraging [106], pollination and seed dispersal [107–110], influence of moonlight [111], sex and reproductive status on the foraging activity [112] are studied in detail. However, the genetic variations within and among populations of *C. sphinx* is not well defined. The lack of genetic information is undoubtedly due, in part, to the difficulties associated with studying them in the wild. The capacity for flight makes bats especially difficult to continuously follow in the wild. In addition, light-tagged animals quickly disappear into dense vegetation making them hard to follow. As a result, data collected by these methods are limited. Direct observation of both sexes are often difficult, therefore genetic analyses may be the only way to obtain reliable data on population structure [31]. An important component required in investigating the population biology of any species is the genetic discrimination of that particular species. This genetic discrimination is the major contributing factor that can help conservation geneticists in evaluating population viability. To provide valuable guidelines for proper conservation and management of *C. sphinx* population, an understanding on genetic diversity is very important.

5.2 Materials and methods

Extensive field trips were carried out to collect *C. sphinx* from different geographical locations in southern Tamil Nadu, India. Bats were captured at the time of emergence from the foliage tents of *P. longifolia* and *B. flabellifer* using a hoop net with an extensible aluminum pole. A small piece of wing membrane from each bat was collected using a sterile biopsy-punch. Tissue samples were obtained from a total of 472 bats from 40 zones. Tissue samples were stored in 70% ethanol at −20°C until DNA extraction [94]. Polymorphism at molecular level was studied by RAPD DNA marker technique. Polymerase chain reaction with 30 arbitrary decamer oligonucleotide primers was applied to the 40 zone samples and to investigate the genetic diversity within and among the populations of *C. sphinx*.

5.3 Results and discussion

Genetic variation is the raw material of evolution and its magnitude is therefore of vital interest in governing the potential of a species to evolve and adapt [96]. The genetic analysis of RAPD markers showed a reasonably high level of diversity. High level of polymorphism was observed in this study which indicates that the genetic base from different zonal population was diverse and extensive. The percentage of polymorphic bands of RAPD was observed to be higher in this species (73.1%). The amount of dispersal and the formation of new social groups are the two factors that strongly affect the genetic structure of the population [113]. Population genetic data from a taxonomically diverse array of social mammals revealed low to moderately high level of genetic differentiation among social groups. This high level of heterozygosity within social groups may be a common feature of mammalian population. The majority of mammalian species exhibit a social system characterized by polygynous-mating and female philopatry [17].

C. sphinx is a polygynous-mating bat and both sexes were found to disperse completely from their natal harems [4]. Moreover, it was observed that the young females either became a member of an already established harem group much earlier when compared to their male counterparts or formed a new harem group of subadult females with an adult male. As a result, the colonies were mainly composed of females which are unrelated or distantly related and with diverse age group [4]. This method of group formation by this species enhances genetic variation. Currently, the high level of genetic diversity can be explained using three factors (i) natal dispersal (ii) formation of new groups and (iii) gene flow between the zones. These are a few such probable reasons for some of the zones to be closely related at the genetic level, although geographically they are from distinct zones of highly distinct locations in Tamil Nadu. This situation can arise in natural populations when there is a possibility of free/random mating and this association between genotypes from contiguous zones may be the result of similar geographical habitat conditions. In addition, recent habitat loss and degradation, which may have led to the concentration of the surviving individuals in the remaining areas, the long generation time and lifespan of the species allowed populations to retain diversity for long periods after habitat loss [17].

Genetic differentiation coefficient of *C. sphinx* from RAPD analysis suggests that the species is of a higher genetic diversity among populations than other bat species [94]. For example, the Brazilian free-tailed bat *Tadarida brasiliensis*, southwestern populations that include those occupying distinct migrational groups show low level of genetic differentiation among populations, even though banding and recapture data suggest low exchange among migratory groups and the inter-colony differences in the bat species are even lower [114]. Similarly, the range of genetic mixing during the seasonal migration of the little red flying-fox *Pteropus scapulatus,* exceeded 3.5 million km^2 [103]. Low degree of differentiation among populations and large amount of gene flow between sub-populations was elaborated using allozymes and RAPD data. A similar result has been reported in gray-headed flying fox *P. poliocephalus* [115].

Genetic studies of migratory bats support high level of gene flow among populations even when separated by large geographical distances (up to 4000 km) [102]. Studying the migratory species using mtDNA markers can further confirm the predicted pattern with little or no genetic structure over broad distance. The individuals of lesser long-nosed bats *Leptonycteris curasoae*, shared identical mtDNA haplotypes when sampled at distances up to 1800 km apart [116]. Similar results have been reported in *P. alecto* [117], *T. brasiliensis* [118], *M. myotis* [119], *Hipposideros speoris* and *Megaderma lyra* [105, 109]. The pattern of population structure and gene flow in species that do not undergo seasonal migration is less clearly known although, in general, gene flow among populations appear more restricted than in migratory species. The gene flow mainly occurs through extra-copulation between the colonies without permanent dispersal from the natal colony [105]. But the distance, availability of mating sites or the recently fragmented population might limit the gene flow. Interestingly, no natal dispersal was found to occur in both the sexes, while extra-colony copulation was observed in most animal species [120].

A greater range of genetic differentiation was identified among the migratory species. Also, a significant correlation between geographic and genetic distances is explained in several species. Extraordinarily, in the Australian ghost bat *Macroderma gigas* the degree of structure was found to be high, with significant correlation between geographic and genetic distances studied using both microsatellites and mtDNA markers [121], similar, results have also been reported in *P. auritus* [30], *M. bechsteinii* [33], *Rhinolophus affinis* [122] and the non-migratory island population of *Eidolon helvum* [123], although as the latter two species were located on islands, gene flow may have been also restricted by sea crossing distance. From such studies, it is apparent that whilst individual colonies within a population may show some genetic heterogeneity due to co-ancestry, little genetic subdivision is apparent, possibly due to low reproductive skew or high levels of dispersal [124]. Moreover, differences in social structure are frequently associated with different mating and dispersal behaviors, which also influence the amount of gene flow among groups and populations [125]. However, not all sedentary species show evidence of population subdivision, even at considerable geographic scales. In particular, genetically effective gene flow appears to occur among populations of vampire bats *D. rotundus* distributed from Mexico to Costa Rica [126]. As discussed above, the molecular studies at inter-population level has verified a greater diversity of population genetic structure within the order.

Seasonal movement is expected to be the main influence among the populations of migratory species because the genetic structure generally appears to be low. However, a wide range of factors including dispersal ability, extrinsic barriers to gene flow and historical events determines the degree of genetic partitioning among population of sedentary species [102]. Dispersal and migration do not essentially equate with the gene flow and hence it is important to consider this factor while accessing the impact of migratory behavior on the genetic structure of bat population. In migratory species, the level genetic structure can be low only when the individual's mate during their migration. Patterns of genetic population structure for both migratory and non-migratory species may resemble if mating and conception in migratory species occur prior to their migration [102]. Gene flow may also be greater than the dispersal capability of individuals of a species which might indicate, provided the population distribution is continuous. For example, radio tracking of individual brown long-eared bats *P. auritus* showed that maximum foraging distances from the summer roost were no greater than 2.8 and 2.2 km for males and females, respectively. Furthermore, this non-migratory species is not thought likely to travel much further at other times of the year [30]. A hierarchical analysis of genetic population structure in *P. auritus* across North-east Scotland identified no genetic differentiation among three adjacent regions when data from colonies within each region were combined. This suggests that colonies across the three regions of North-east Scotland form a continuously distributed population, within which genes move *via* a `stepping stone' model [30].

In our study, the maximum similarity was observed as many zones were closer to each other. Therefore, when populations remain closer, the gene flow is expected to be greater. As a result, the nearby populations should remain more similar at neutral loci. This relationship is referred as the method of isolation by distance and serves as the stepping stone model of gene flow [127]. However, the distance between populations and the nature of the surrounding landscape between population are the two factors on which the level of gene flow depends [128]. These findings support that *C. sphinx* is not known to undergo seasonal migrations. Moreover, it is a common plant-visiting bat that occurs throughout India and much of mainland Southeast Asia [1]. Our results showing the high genetic variations in *C. sphinx* population is not surprising because, the distribution of these bats is continuous and the level of gene flow is also high. Similarly, study has been carried out among *C. sphinx* of the Indian subcontinent which suggests high gene flow and equilibrium population dynamics [129]. Thus, for the long-term persistence of *C. sphinx* populations, maintaining the gene flow is considered as a key factor.

5.4 Conclusion

Our study deals with the genetic diversity in natural population of *C. sphinx* at the molecular level. We concluded that *C. sphinx* population maintains high levels of genetic variability despite of increase in fragmentation of their habitat. Though this may be beneficial factor for the conservation of these bat species, some caution should be observed. The results suggest that bats move rather freely between zones and current bat populations may continue to decrease in many of the habitats investigated. Furthermore, *C. sphinx* is still a relatively widespread species; it has suffered dramatic population declines during the past several years. Using coalescent based Bayesian analysis, a significant demographic contraction was found to be evident among a large sample of *C. sphinx* genotypes [129], which were one of the eight localities included in the Indian latitudinal study [130]. These results suggest that Indian *C. sphinx* is strongly associated with open habitat [90]. In addition, our direct observation and mark recapture data show a gradual decline of natural populations of *C. sphinx*. However, this study provides baseline genetic information for future studies. To look at the long-term effects of human induced habitat fragmentation and degradation on genetic diversity and structure, microsatellite and mitochondrial DNA variation should be reassessed among this species. It can be concluded that RAPD analysis revealed high levels of genetic polymorphism and differentiation might play a role in the dynamic evolution of *C. sphinx* in southern India. These results would help in developing an effective and meaningful method in conservation of this species. Future studies of Old-World fruit bats from these areas will be of great bio-geographic and evolutionary interest.

Author details

Thangavel Karuppudurai* and Steffi Christiane Ramesh
Department of Animal Behaviour and Physiology, School of Biological Sciences, Madurai Kamaraj University, Madurai, Tamil Nadu, India

*Address all correspondence to: tkdurai@gmail.com

References

[1] Storz JF, Kunz TH. *Cynopterus sphinx*. Mammal Species. 1999; 613: 1-8.

[2] Balasingh J, Koilraj J, Kunz TH. Tent construction by the short-nosed fruit bat *Cynopterus sphinx* (Chiroptera: Pteropodidae) in southern India. Ethology. 1995; 100: 210-229.

[3] Storz JF, Balasingh J, Nathan PT, Emmanuel K, Kunz TH. Dispersion and site fidelity in a tent-roosting population of the short-nosed fruit bat (*Cynopterus sphinx*) in southern India. Journal of Tropical Ecology. 2000a; 16: 117-131.

[4] Storz JF, Bhat HR, Kunz TH. Social structure of a polygynous tent-making bat, *Cynopterus sphinx* (Megachiroptera). Journal of Zoology. 2000b; 251: 151-165.

[5] Balasingh J, Isaac SS, Subbaraj R. Tent-roosting by the frugivorous bat, *Cynopterus sphinx* (Vahl 1797) in southern India. Current Science. 1993; 65: 418.

[6] Bhat HR, Kunz TH. Altered flower/fruit clusters of the kitul palm used as roosts by the short-nosed fruit bat, *Cynopterus sphinx* (Chiroptera: Pteropodidae). Journal of Zoology. 1995; 235: 597-604.

[7] Kunz TH. Roosting ecology of bats. In: Kunz TH, editor. Ecology of Bats. New York: Plenum Press; 1982. p. 1-50.

[8] Gopukumar N, Karuppudurai T, Nathan PT, Sripathi K, Arivarignan G, Balasingh J. Solitary adult males in a polygynous-mating bat (*Cynopterus sphinx*): a forced option or a strategy? Journal of Mammalogy. 2005; 86: 281-286.

[9] Karuppudurai T, Gopukumar N, Sripathi K. Solitary or nonterritorial adult males in bats are "Making the best of a bad job"? Bat Net. 2006; 7: 30-33.

[10] Karuppudurai T, Sripathi K, Gopukumar N, Elangovan V, Arivarignan G. Transition of nonharem male to harem male status in the short-nosed fruit bat *Cynopterus sphinx*. Mammalian Biology. 2008; 73: 138-146.

[11] Krishna A, Dominic CJ Reproduction in the female short-nosed fruit bat *Cynopterus sphinx* (Vahl). Period Biology. 1983; 85: 23-30.

[12] Sandhu S. Breeding biology of the Indian fruit bat, *Cynopterus sphinx* (Vahl) in central India. Journal of the Bombay Natural History Society. 1984; 81: 600-611.

[13] Ramakrishna PA. Post-partum oestrus in the Indian short-nosed fruit bat, *Cynopterus sphinx sphinx* (Vahl.). Current Science. 1947; 16: 186.

[14] Gopalakrishna A. Gestation period in some Indian bats. Journal of the Bombay Natural History Society. 1969; 66: 317-322.

[15] Bhat HR, Sreenivasan MA. Records of bats in Kyasanur forest disease area and environs in Karnataka state, India, with ecological notes. Mammalia. 1990; 54: 69-106.

[16] Elangovan V, Priya EYS, Raghuram H, Marimuthu G. Postnatal development in the Indian short-nosed fruit bat *Cynopterus sphinx*: growth rate and age estimation. Acta Chiropterologica. 2003; 5: 107-116.

[17] Greenwood PJ. Mating systems, philopatry and dispersal in birds and mammals. Animal Behaviour. 1980; 28: 1140-1162.

[18] Clobert J, Nichols JD, Danchin E, Dhondt A. Dispersal. Oxford: Oxford University Press; 2001. 452 p.

[19] Baillie SR, Sutherland WJ, Freeman SN, Gregory RD, Paradis E.

Consequences of large-scale processes for the conservation of bird populations. Journal of Applied Ecology. 2000; 37: 88-102.

[20] McCracken GF, Bradbury JW. Paternity and Genetic Heterogeneity in the Polygynous Bat, Phyllostomus hastatus. Science. 1977; 198: 303-306.

[21] Handley Jr. CO, Wilson DE, Gardner AL. Demography and natural history of the common fruit bat, *Artibeus jamaicensis*, on Barro Colorado Island, Panama. Washington, DC: Smithsonian Institution Press; 1991. 511 p.

[22] Castella V, Ruedi M, Excoffier L. Contrasted patterns of mitochondrial and nuclear structure among nursery colonies of the bat *Myotis myotis*. Journal of Evolutionary Biology. 2001; 14: 708-720.

[23] Kerth G, Morf L. Behavioural and Genetic Data Suggest that Bechstein's Bats Predominantly Mate Outside the Breeding Habitat. Ethology. 2004; 110: 987-999.

[24] McCracken GF, Bradbury JW. Social organization and kinship in the polygynous bat *Phyllostomus hastatus*. Behavioural Ecology and Sociobiology. 1981; 8: 11-34.

[25] Fleming TH. The short-tailed fruit bat: a study in plant-animal interactions. Chicago: University of Chicago Press; 1988.

[26] Bain JR, Humphrey SR. Social organization and biased primary sex ratio of the evening bat, *Nycticeius humeralis*. Florida Academy of Sciences. 1986; 49: 22-31.

[27] Dechmann DKN, Kalko EKV, Kerth G. All-offspring dispersal in a tropical mammal with resource defense polygyny. Behavioural Ecology and Sociobiology. 2007; 61: 1219-1228.

[28] Wilkinson GS, Chapman AM. Length and sequence variation in evening bat D-Loop mtDNA. Genetics. 1991; 128: 607-617.

[29] Wilkinson GS. Altruism and co-operation in bats. In: Fenton MB, Racey P, Rayner JM, editors. Recent advances in the study of Bats. Cambridge: Cambridge University Press; 1987. p. 299-323.

[30] Burland TM, Barratt EM, Beaumont MA, Racey PA. Population genetic structure and gene flow in a gleaning bat, *Plecotus auritus*. Proceedings of the *Royal Society* of *London*. Series *B*, Biological Sciences. 1999; 266: 975-980.

[31] Burland TM, Barratt EM, Nichols RA, Racey PA. Mating patterns, relatedness and the basis of natal philopatry in the brown long-eared bat, *Plecotus auritus*. Molecular Ecology. 2001; 10: 1309-1321.

[32] Kerth G, König B. Fission, fusion and nonrandom associations in female Bechstein's bats (*Myotis bechsteinii*). Behaviour. 1999; 136: 1187-1202.

[33] Kerth G, Mayer F, König B. Mitochondrial DNA (mtDNA) reveals that female Bechstein's bats live in closed societies. Molecular Ecology. 2000; 9: 793-800.

[34] Kerth G, Wagner M, König B. Roosting together, foraging apart: information transfer about food is unlikely to explain sociality in female Bechstein's bats (*Myotis bechsteinii*). Behavioural Ecology and Sociobiology. 2001; 50: 283-291.

[35] Rossiter SJ, Jones G, Ransome RD, Barratt EM. Relatedness structure and kin-biased foraging in the greater horseshoe bat (Rhinolophus ferrumequinum). Behavioural Ecology and Sociobiology. 2002; 51: 510-518.

[36] Arnold BD. Population structure and sex-biased dispersal in the forest dwelling Vespertilionid bat, *Myotis septentrionalis*. American Midland Naturalist. 2007; 157: 374-384.

[37] Clutton-Brock TH. Female transfer and inbreeding avoidance in social mammals. Nature. 1989a; 337: 70-72.

[38] Usman K. Some aspect of the population dynamics of the bat, *Rhinopoma hardwickei*, in a cave system. Journal of the Bombay Natural History Society. 1984; 83: 120-129.

[39] Entwistle AC, Racey PA, Speakman JR. Social and population structure of a gleaning bat, *Plecotus auritus*. Journal of Zoology. 2000; 252: 11-17.

[40] Palmeirim JM, Rodrigues L. Dispersal and philopatry in colonial animals: the case of *Miniopterus schreibersii*. Symposia of the Zoological Society of London. 1995; 67: 219-231.

[41] Cheng HC, Lee LL. Temporal variations in the size and composition of formosan Leaf-nosed bat (*Hipposideros terasensis*) colonies in central Taiwan. Zoological Studies. 2004; 43: 787-794.

[42] Sandhu S, Gopalakrishna A. Some observations on the breeding biology of the Indian fruit bat, *Cynopterus sphinx* (Vahl) in central India. Current Science. 1984; 53: 135-142.

[43] Clutton-Brock TH. Mammalian mating systems. Proceedings of the *Royal Society* of *London*. Series *B*, Biological Sciences. 1989b; 236: 339-372.

[44] Bradbury JW, Vehrencamp SL. Social organizations and foraging in emballonurid bats, II. A model for the determination of group size. Behavioral *Ecology* and *Sociobiology.* 1977; 1: 383-404.

[45] Christy JH. Female choice in the resource-defense mating system of the sand fiddler crab, *Uca pugilator*. Behavioural Ecology and Sociobiology. 1983; 12: 169-180.

[46] McCracken GF, Wilkinson GS. Bat mating systems. In: Krutzsch PH, Creighton EG, editors. Reproductive biology of Bats. New York: Academic Press; 2000. p. 321-362.

[47] Dechmann DKN, Kalko EKV, König B, Kerth G. Mating system of a Neotropical roost-making bat: the white-throated, round-eared bat, *Lophostoma silvicolum* (Chiroptera: Phyllostomidae). Behavioural Ecology and Sociobiology. 2005; 58 316-325.

[48] Altringham JD. Mating behaviour. In: Altringham JD, editor. Bats biology and behaviour. Oxford: Oxford University Press; 1996. p. 147-154.

[49] Timm RM, Clauson BL. A roof over their feet. Natural History. 1990; 3: 55-58.

[50] Timm RM, Lewis SE. Tent construction and use by *Uroderma bilobatum* in coconut palms (*Cocos nucifera*) in Costa Rica. Bulletin of the American Museum of Natural History. 1991; 206: 251-260.

[51] Brooke AP. Tent construction and social organization in *Vampyressa nymphaea* (Chiroptera: Phyllostomidae) in Costa Rica. Journal of Tropical Ecology. 1987; 3: 171-175.

[52] Brooke AP. Tent selection, roosting ecology and social organization of the tent-making bat, *Ectophylla alba* , in Costa Rica. Journal of Zoology. 1990; 221: 11-19.

[53] Morrison DW. Apparent Male Defense of Tree Hollows in the Fruit Bat, Artibeus jamaicensis. Journal of Mammalogy. 1979; 60: 11-15.

[54] Morrison DW, Morrison SH. Economics of harem maintenance by a

Neotropical bat. Ecology. 1981; 62: 864-866.

[55] Kunz TH, McCracken GF. Tents and harems: apparent defence of foliage roosts by tent-making bats. Journal of Tropical Ecology. 1996; 12: 121-137.

[56] Kunz TH, Lumsden LF. Ecology of cavity and foliage roosting bats. In: Kunz TH, Fenton MB, editors. Bat ecology. Chicago: University of Chicago Press; 2003. p. 3-90.

[57] Kunz TH, Fujita MS, Brooke AP, McCracken GF. Convergence in tent architecture and tent-making behavior among neotropical and paleotropical bats. Journal of Mammalian Evolution. 1994; 2: 57-78.

[58] Storz JF, Bhat HR, Kunz TH. Genetic consequences of polygyny and social structure in an Indian fruit bat, *Cynopterus sphinx*. II. Variance in male mating success and effective population size. Evolution. 2001b; 55: 1224-1232.

[59] Ortega J, Arita HT. Structure and social dynamics of harem groups in *Artibeus jamaicensis* (Chiroptera: Phyllostomidae). Journal of Mammalogy. 1999; 80: 1173-1185.

[60] Ortega J, Arita HT. Defence of Females by Dominant Males of *Artibeus jamaicensis* (Chiroptera: Phyllostomidae). Ethology. 2000; 106: 395-407.

[61] Kunz TH, Robson SK, Nagy KA. Economics of harem maintenance in the greater spear-nosed bat, *Phyllostomus hastatus*. Journal of Mammalogy. 1998; 79: 23-30.

[62] Gopukumar N, Manikandan M, Arivarignan G. Roosting patterns in a captive colony of short-nosed fruit bat *Cynopterus sphinx* (Vahl). Indian Journal of Experimental Biology. 2002; 40: 1187-1190.

[63] Kunz TH, August PV, Burnett CD. Harem social organization in cave roosting *Artibeus jamaicensis* (Chiroptera: Phyllostomidae). Biotropica. 1983; 15: 133-138.

[64] Voigt CC. Individual variation in perfume blending in male greater sac-winged bats. Animal Behaviour. 2002; 63: 907-913.

[65] McDonald DB, Potts WK. Cooperative display and relatedness among males in a lek-mating bird. Science. 1994; 266: 1030-1032.

[66] Emlen ST, Oring LW. Predicting family dynamics in social vertebrates. In: Krebs JR, Davies NB, editors. Behavioural ecology: an evolutionary approach. Oxford: Blackwell Publishing; 1997. p. 228-253.

[67] Danforth BN. The morphology and behavior of dimorphic males in *Perdita portalis* (Hymenoptera: Andrenidae). Behavioural Ecology and Sociobiology. 1991; 29: 235-247.

[68] Martin E, Taborsky M. Alternative male mating tactics in a cichlid, *Pelvicachromis pulcher*: a comparison of reproductive effort and success. Behavioural Ecology and Sociobiology. 1997; 41: 311-319.

[69] Gross MR. Alternative reproductive strategies and tactics: diversity within sexes. Trends in Ecology and Evolution. 1996; 11: 92-98.

[70] Stockley P, Searle JB, Macdonald DW, Jones CS. Alternative reproductive tactics in male common shrews: relationships between mate-searching behaviour, sperm production, and reproductive success as revealed by DNA fingerprinting. Behavioural Ecology and Sociobiology. 1994; 34: 71-78.

[71] Dewsbury DA. Dominance rank, copulatory behavior, and differential

reproduction. Quarterly *Review* of *Biology*. 1982; 57: 135-159.

[72] Perrill SA, Gerhardt HC, Daniel R. Sexual Parasitism in the Green Tree Frog (Hyla cinerea). Science. 1978; 200: 1179-1180.

[73] Wirtz P. Territory holders, satellite males and bachelor males in a high-density population of water buck (*Kobus ellipsiprymnus*) and their associations with conspecifics. Zeitschrift für Tierpsychologie. 1982; 58: 277-300.

[74] Clutton-Brock TH, Albon SD, Gibson RM, Guinness FE. The logical stag: adaptive aspects of fighting in red deer (*Cervus elaphus* L.). Animal Behaviour. 1979; 27: 211-225.

[75] Gross MR. Sneakers, satellites and parentals: polymorphic mating strategies in North American sunfishes. Zeitschrift für Tierpsychologie. 1982; 60: 1-26.

[76] Wilkinson GS. The social organization of the common vampire bat. II. Mating system, genetic structure and relatedness. Behavioural Ecology and Sociobiology. 1985; 17: 123-134.

[77] Heckel G, Voigt GG, Mayer F, von Helversen O. Extra harem paternity in the white-lined bat *Saccopteryx bilineata* (Emballonuridae). Behaviour. 1999; 136: 1173-1185.

[78] Ortega J, Maldonado JE, Wilkinson GS, Arita HT, Fleischer RC. Male dominance, paternity, and relatedness in the Jamaican fruit-eating bat (*Artibeus jamaicensis*). Molecular Ecology. 2003; 12: 2409-2415.

[79] Heckel G, von Helversen O. Genetic mating system and the significance of harem associations in the bat *Saccopteryx bilineata*. Molecular Ecology. 2003; 12: 219-227.

[80] Lewis SE. Roost fidelity of bats: a review. Journal of Mammalogy. 1995; 76: 481-496.

[81] Zhan A. Individual migration between colonies of Greater mouse-eared bats (*Myotis myotis*) in Upper Bavaria. Zeitschrift für *Säugetierkunde*. 1998; 63: 321-328.

[82] Rossiter SJ, Jones G, Ransome RD, Barratt EM. Parentage, reproductive success and breeding behaviour in the greater horseshoe bat (*Rhinolophus ferrumequinum*). Proceedings Biological Sciences. 2000b; 267: 545-551.

[83] Cryan PM, Bogan MA, Yanega GM. Roosting habits of four bat species in the Black Hills of South Dakota. Acta Chiropterologica. 2001; 3: 43-52.

[84] Willis CKR, Brigham RM. Roost switching, roost sharing and social cohesion: forest-dwelling big brown bats, *Eptesicus fuscus*, conform to the fission-fusion model. Animal Behaviour. 2004; 68: 495-505.

[85] Eby P. Seasonal movements of grey-headed flying-foxes, *Pteropus poliocephalus* (Chiroptera: Pteropodidae), from two maternity camps in northern New South Wales. Wildlife Research. 1991; 18: 547-559.

[86] Sreenivasan MA, Bhat HR, Geevarghese G. Observations on the reproductive cycle of *Cynopterus sphinx sphinx* Vahl, 1797 (Chiroptera: Pteropidae). Journal of Mammalogy. 1974; 55: 200-202.

[87] Nathan PT, Swami Doss DP, Isaac SS, Balasingh J, Rajan KE, Gopukumar N, Subbaraj R. Mist net capture and field observation on the short-nosed fruit bat (Chiroptera: Pteropodidae) *Cynopterus sphinx*. Journal of the Bombay Natural History Society. 2001; 98: 373-378.

[88] Sandhu S. Observations on the reproduction and associated phenomena in the male fruit bat *Cynopterus sphinx* (Vahl) in Central India. Journal of the Bombay Natural History Society. 1988; 85: 135-142.

[89] Hodgkison R, Balding ST, Akbar Z, Kunz TH. Roosting ecology and social organization of the spotted-winged fruit bat, *Balionycteris maculata* (Chiroptera: Pteropodidae), in a Malaysian lowland dipterocarp forest. Journal of Tropical Ecology. 2003; 19: 667-676.

[90] Storz JF, Balasingh J, Bhat HR, Nathan PT, Doss DPS, Prakash AA, Kunz TH. Clinal variation in body size and sexual dimorphism in an Indian fruit bat, *Cynopterus sphinx* (Chiroptera: Pteropodidae). Biological Journal of the Linnean Society. 2001a; 72: 17-31.

[91] Emlen ST, Oring LW. Ecology, sexual selection and the evolution of mating systems. Science. 1977; 197: 215-223.

[92] Caro TM, Bateson P. Organization and ontogeny of alternative tactics. Animal Behaviour. 1986. *34;* 1483-1499.

[93] Karuppudurai T, Sripathi K. Molecular genetic analysis of male alternative strategy and reproductive success in the polygynous mating bat *Cynopterus sphinx*. Journal of Cell and Animal Biology. 2010; 4: 042-050.

[94] Karuppudurai T, Sripathi K, Gopukumar N, Elangovan V, Marimuthu G. Genetic diversity within and among populations of the Indian short-nosed fruit bat *Cynopterus sphinx* assessed through RAPD analysis. Current Science. 2007; 93: 942-950.

[95] Heckel G, von Helversen O. Male tactics and reproductive success in the harem polygynous bat *Saccopteryx bilineata*. Behavioural Ecology. 2002; 13: 750-756.

[96] Frankham R. Quantitative genetics in conservation biology. Genetics Research. 1999; 74: 237-244.

[97] Blois JL, Arbogast BS. Conservation genetics of the Sonoma tree vole (*Arborimus pomo*) based on mitochondrial and amplified fragment length polymorphism markers. Journal of Mammalogy. 2006; 87: 950-960.

[98] Avise JC, Alisauskas RT, Nelson WS, Ankney CD. Matriarchal population genetic structure in an avian species with female natal philopatry. Evolution. 1992; 46: 1084-1096.

[99] Hewitt GM, Butlin RK. Causes and consequences of population structure. In: Krebs JR, Davies NB, editors. Behavioural ecology: an evolutionary approach. Oxford: Blackwell Publishing Science Ltd.; 1997. p. 203-277.

[100] Bohonak AJ. Dispersal, gene flow, and population structure. Quarterly *Review* of *Biology*. 1999; 74: 21-45.

[101] Avise JC. Molecular markers, natural history and evolution. Chapman and Hall, New York; 1994. 511 p.

[102] Burland TM, Wilmer JW. Seeing in the dark: molecular approaches to the study of bat populations. Biological Review. 2001; 76: 389-409.

[103] Sinclair EA, Webb NJ, Marchant AD, Tidemann CR. Genetic variation in the little red flying-fox *Pteropus scapulatus* (Chiroptera: Pteropodidae): implications for management. Biological Conservation. 1996; 76: 45-50.

[104] Rossiter SJ, Jones G, Ransome RD, Barratt EM. Genetic variation and population structure in the endangered greater horseshoe bat *Rhinolophus ferrumequinum*. Molecular Ecology. 2000a; 9: 1131-1135.

[105] Emmanuvel Rajan K, Marimuthu G. A preliminary examination of genetic diversity in the Indian false vampire bat *Megaderma lyra*. *Animal Biodiversity* and *Conservation*. 2006; 29: 109-115.

[106] Gopukumar N, Elangovan V, Sripathi K, Marimuthu G, Subbaraj R.

Foraging behaviour of the Indian short-nosed fruit bat *Cynopterus sphinx*. Zeitschrift für *Säugetierkunde*. 1999; 64: 187-191.

[107] Elangovan V, Marimuthu G, Kunz TH. Nectar feeding behaviour in the short-nosed fruit bat *Cynopterus sphinx* (Pteropodidae). Acta Chiropterologica. 2000; 2: 1-5.

[108] Singaravelan N, Marimuthu G. Nectar feeding and pollen carrying from *Ceiba pentandra* by pteropodid bats. Journal of Mammalogy. 2004; 85: 1-7

[109] Emmanuvel Rajan K, Marimuthu G. Genetic diversity among populations of the microchiropteran bat *Hipposideros speoris* based on RAPD analysis. Mammalian Biology. 2000; 65: 301-306.

[110] Nathan PT, Raghuram H, Elangovan V, Karuppudurai T, Marimuthu G. Bat pollination of kapok tree, *Ceiba pentandra*. Current Science. 2005; 88: 1679-1681.

[111] Singaravelan N, Marimuthu G. Moonlight inhibits and lunar eclipse enhances foraging activity of fruit bats in an orchard. Current Science. 2002; 82: 1020-1022.

[112] Marimuthu G, Rajan KE, Koilraj AJ, Isaac SS, Balasingh J. Observations on the foraging behavior of a tent roosting megachiropteran bat *Cynopterus sphinx*. Biotropica. 1998; 30: 321-324.

[113] Perrin N, Mazalov V. Local competition, inbreeding, and the evolution of sex-biased dispersal. The American Naturalist. 2000; 155: 116-127.

[114] Svoboda PL, Choate JR, Chesser RK. Genetic relationships among southwestern populations of the Brazilian Free-tailed bat. Journal of Mammalogy. 1985; 66: 444-450.

[115] Webb NJ, Tidemann CR. Hybridisation between black (*Pteropus alecto*) and grey-headed (*P. Poliocephalus*) flying-foxes (megachiroptera: Pteropodidae). Australian Mammalogy. 1995; 18: 19-26.

[116] Wilkinson GS, Fleming TH. Migration and evolution of lesser long-nosed bats *Leptonycteris curasoae*, inferred from mitochondrial DNA. Molecular Ecology. 1996; 5: 329-339,

[117] Webb NJ, Tidemann CR. Mobility of Australian flying-foxes, *Pteropus spp*. (Megachiroptera): evidence from genetic variation. Proceedings Biological Sciences. 1996; 263: 497-502.

[118] McCracken GF, Gassel MF. Genetic structure in migratory and nonmigratory populations of Brazilian free-tailed bats. Journal of Mammalogy. 1997; 78: 348-357.

[119] Castella V, Ruedi M, Excoffier L, Ibáñez C, Arlettaz R, Hausser J. Is the Gibraltar Strait a barrier to gene flow for the bat *Myotis myotis* (Chiroptera: Vespertilionidae)? Molecular Ecology. 2000; 9: 1761-1772.

[120] Amos B, Schlötterer C, Tautz D. Social structure of pilot whales revealed by analytical DNA profiling. Science. 1993; 260: 670-672.

[121] Wilmer JW, Hall L, Barratt E, Moritz C. Genetic structure and male-mediated gene flow in the ghost bat (*Macroderma gigas*). Evolution. 1999; 53: 1582-1591.

[122] Maharadatunkamsi, Hisheh S, Kitchener DJ, Schmitt LH. Genetic and morphometric diversity in Wallacea: geographical patterning in the horse shoe bat, *Rhinolophus affinis*. Journal of Biogeography. 2000; 27: 193-201.

[123] Juste J, Ibáñez C, Machordom A. Morphological and allozyme variation of *Eidolon helvum* (Mammalia:

megachiroptera) in the islands of the Gulf of Guinea. Biological Journal of the Linnean Society. 2000; 71: 359-378.

[124] McCracken GF. Genetic structure of bat social groups. In: Fenton MB, Racey P, Rayner JMV, editors. Recent advances in the study of Bats. Cambridge: Cambridge University Press; 1987. p. 281-298.

[125] Nunney L. The influence of mating system and overlapping generations on effective population size. Evolution. 1993; 47: 1329-1341.

[126] Honeycutt RL, Greenbaum IF, Baker RJ, Sarich VM. Molecular evolution of vampire bats. Journal of Mammalogy. 1981; 62: 805-811.

[127] Kimura M, Weiss GH. The stepping stone model of population structure and the decrease of genetic correlation with distance. Genetics. 1964; 49: 561-576.

[128] Gibbs JP. Demography versus habitat fragmentation as determinants of genetic variation in wild populations. Biological Conservation. 2001; 100: 15-20.

[129] Storz JF. Contrasting patterns of divergence in quantitative traits and neutral DNA markers: analysis of clinal variation. Molecular Ecology. 2002; 11: 2537-2551.

[130] Storz JF, Beaumont MA. Testing for genetic evidence of population expansion and contraction: an empirical analysis of microsatellite DNA variation using a hierarchical Bayesian model. Evolution. 2002; 56: 154-166.

Section 6

Roost-Tier Preference

Chapter 7

Roost-Tier Preference in Roost-Trees: A Case Study in the Bats *Pteropus giganteus*

Susanta Mallick, Asif Hossain and Srimanta Kumar Raut

Abstract

The Indian flying foxes *Pteropus giganteus* are habituated to spend the day hours roosting in suitable roost trees. They are seen hanging here and there in a roost tree. It is not known whether they have preferred roost sites rather hanging spots in the concerned roost tree. To testify the said hypothesis we selected two roost trees, *Albizia lebbeck* and *Tamarindus indica* locating at distant places (75 km apart) in the arid zone of West Bengal, India during the period of last ten years. It is revealed that *P. giganteus* preferred branches of the roost tree which are locating in the mid-tier of tree. But depending upon the situations the less preferred sites are not spared as these sites are used by the late comers. Statistical tests following application of one-way ANOVA justified significant effect of the roost branch on the abundance of bat population ($P<0.05$), abundance of bats in the roost branches is highly correlated in respect to the study years ($r=0.96$) is also justified from the study of normality distribution plot, and the results of GLMM strongly support the hypothesis irrespective of the variables, that is branches of the roost tree and the year of observations ($P = 0.0$).

Keywords: *Pteropus giganteus* bats, roost tree, roost branches, roost-tier preference

1. Introduction

Bats roost mostly in caves and trees. These roost sites are degrading day by day because of unpredictable human activities [1] especially due to destruction of roost trees at large [2–11]. Customarily tree roosting bats select certain aged trees having well developed canopy area [9, 12–14] it is not clearly known which part of a roost tree is preferred by the bats and why? Or there exists no discretion in respect to roost site selection in a tree.

The Indian flying foxes *Pteropus giganteus* [15] found in India, Bangladesh, Nepal, Bhutan, China, Maldives, Myanmar, Pakistan and Sri Lanka [16]. These frugivorous flying mammals are habituated to spend the day hours at the roost sites specially in selective trees in open spaces [2, 4, 6, 13, 17, 18]. Mostly; these bats select big, well branched and leafy trees for roosting. As there exists many branches of a roost tree and usually the branches are gradually smaller in length with increasing height of the tree we aimed to study the preferential sites, if any, the bats considered for hanging. Accordingly, we selected two roost trees *Albizia lebbeck* (L.) Benth and *Tamarindus indica* (L.) locating at distant places in the village area. The results, we obtained are very much impacted by the intra-specific competition even if these bats are socially well organized and the members of a colony are guided by the social bindings to carry out allotted duties assigned for the well being of the colony members.

2. Materials and methods

We selected two roost-trees for the proposed studies. Of the two, one, the silk flower tree *Albizia lebbeck* (16.15 m in height, umbrella shaped crown, 1637.96 m^2 in canopy area with 1.73 m in diameter) with five main branches bearing thin foliages. This tree is deciduous in nature. It is located in the village Joteghanashyam (22°31′10.0"N, 87°50′19.2″ E) of Paschim Medinipur district. The second roost-tree *Tamarindus indica* (17.67 m in height, umbrella shaped crown, 1960.79 m^2 in canopy area with 3.33 m in diameter) with 14 main branches bearing thick foliages. This is evergreen in nature. It is located in the village Simla (23°22′44.20"N, 86°38′47.02″ E) of Purulia district, 75 km west of silk flower tree; these two districts are locating in the arid zone of state of West Bengal, India.

In *A. lebbeck*, of the five main branches, the longest one was 15 m in length while the smallest one was confined to 7 m. In contrast, in *T. indica* the longest branch was 13 m and the shortest one was 3 m in length. In both cases such measurements were taken on the last sampling dates. Irrespective of roost-trees there were numerous short sub-branches at certain points along the extended parts of the main stem from the point of emergence of the main stem body.

These branches from the lower to upper part of the tree were marked as L_1, L_2, L_3 and so on depending upon the number of branches occurring successively up to the top of the tree. Thus in case of *A. lebbeck* branches were numbered as L_1 to L_5 and for *T. indica* the same was ranged from L_1 to L_{14}. The lower most branch of *A. lebbeck* and *T. indica* was 5.79 m and 1.78 m above the ground respectively. We counted the number of bats hanged in respect to the marked branch including the sub-branches of the same and the data were recorded at monthly interval. We used binocular as and when necessary to locate the bats to avoid any kind of ambiguity in counting of the bats. The counting was initiated on 25 April 2011 and continued up to 30 March 2021 at Joteghanashyam, and from 19 January 2015 to 23 December 2020 at Simla.

3. Statistical analysis of the data

Data collected were pooled together to estimate the average number of bat individuals selected the specific branch of the roost trees, irrespective of months of the study years as well as the standard error (SE) values. One-way ANOVA was applied to justify whether the branches have significant effect in selecting the same as roost sites by the bats. Normal probability Plot of PAST Software was used to ascertain and justify the normal distribution of the roosting bats in different branches of the roost trees. GLMM was applied to testify the proposed hypothesis by determining the overall significance levels ($P<0.05$).

4. Results

It is revealed that the roosting abundance of *P. giganteus* varied from 73.31±6.52 to 217.19±20.88 in *A. lebbeck* at Joteghanashyam (**Figure 1**) and

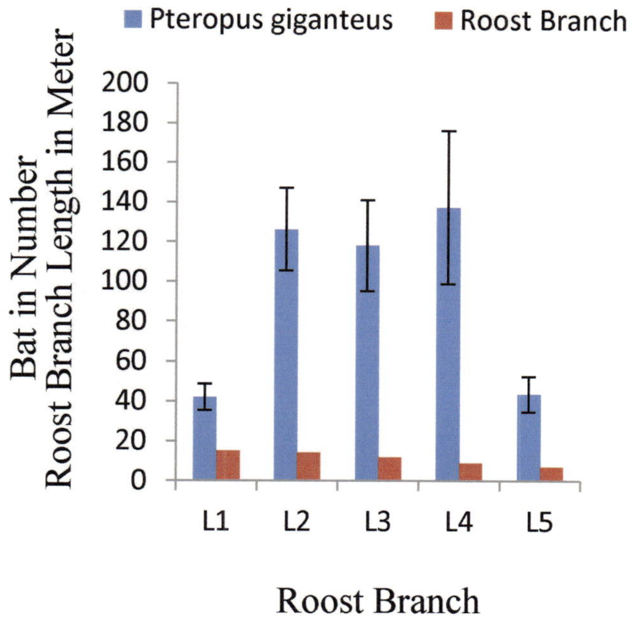

Figure 1.
Mean (±SE) number of P. giganteus *bats used the branches of the roost tree* A. lebbeck *daily during 2011–2021 study periods at Joteghanashyam.*

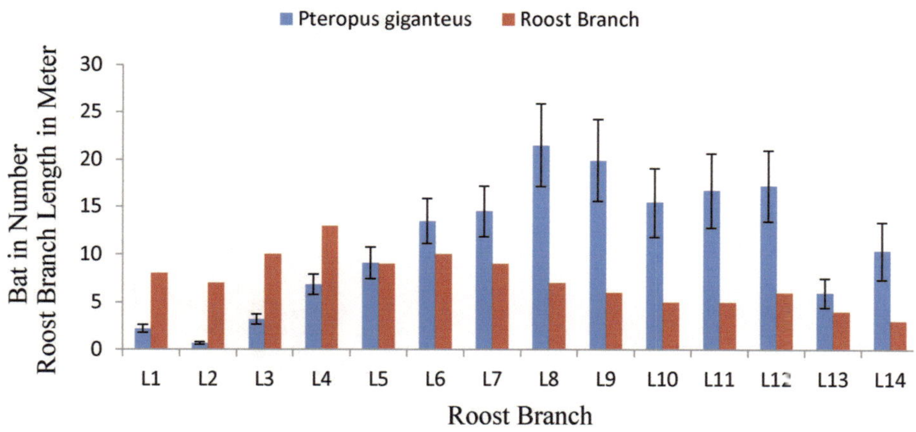

Figure 2.
Mean (±SE) number of P. giganteus *bats used the branches of the roost tree* T. indica *daily during 2015–2020 study periods at Simla.*

1.12±0.13 to 37.64±4.39 in *T. indica* at Simla (**Figure 2**) per roost branch. Results of ANOVA test clearly indicate that there exists significant differences in selection of the roost sites by *P. giganteus* in *A. lebbeck* (df=14, F=5.71, P=0.00, N=540) and *T. indica* (df=12, F=2.05, P=0.00, N=373). From the normality distribution plot (**Figures 3** and **4**) it is evident that there exists significant correlation between abundance of roosting bat population and the study years. GLMM studies (**Table 1**) confirmed that the roost sites occupied by *P. giganteus*, in the branches of both the roost trees (except one branch L_1 in *A. lebbeck* and two branches L_2 and L_3 in *T. indica*) are undoubtedly preferred sites for roosting (P=0.0) depending upon the probability of availability of these sites upon their time of return to the roost tree.

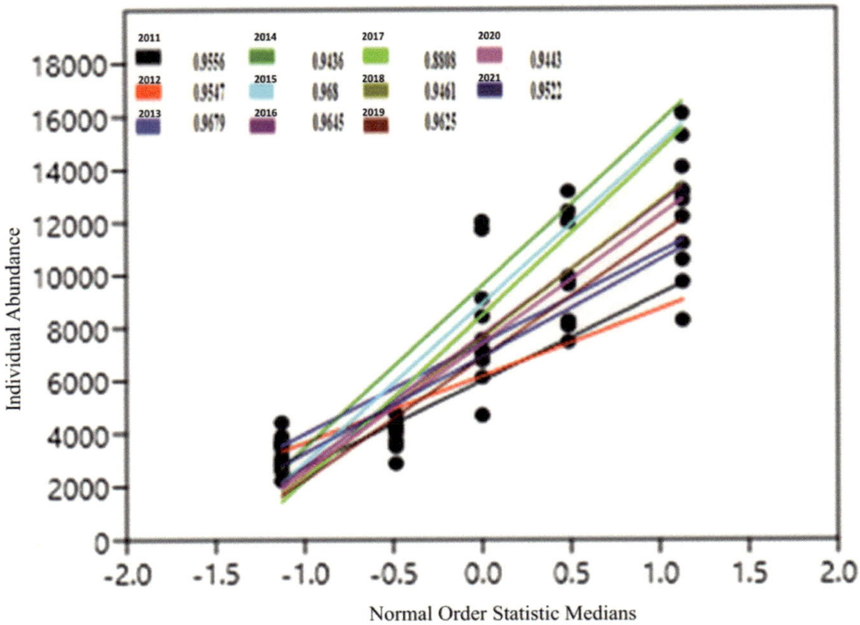

Figure 3.
Normal probability distribution of P. giganteus *in* A. lebbeck *roost tree during study years (2011–2021) at Joteghanashyam.*

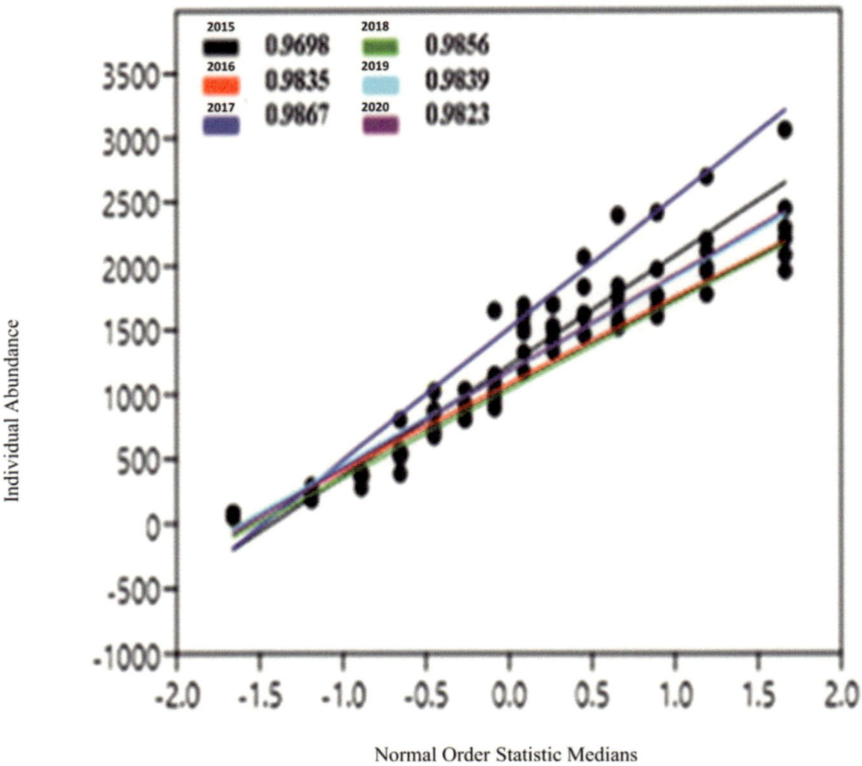

Figure 4.
Normal probability distribution of P. giganteus *in* T. indica *roost tree during study years (2015–2020) at Simla.*

Albizia lebbeck				***Tamarindus indica***			
Variables	**Estimate**	**T-value**	**P**	**Variables**	**Estimate**	**T-value**	**P**
(Intercept)	3598.7273	5.1964979	0	(Intercept)	239.3333	2.426218	0
BLL2	7063.3636	7.3529123	0	BLL2	−169.3333	1.614338	>0.05
BLL3	6005.3636	6.2515416	0	BLL3	119.5000	1.139252	>0.05
BLL4	6528.5455	6.7961703	0	BLL4	557.1667	5.311744	0
BLL5	156.6364	0.1630574	>0.05	BLL5	780.1667	7.437713	0
				BLL6	1285.0000	12.250537	0
				BLL7	1383.3333	13.187997	0
				BLL8	2100.5000	20.025100	0
				BLL9	1881.3333	17.935677	0
				BLL10	1361.6667	12.981438	0
				BLL11	1495.0000	14.252571	0
				BLL12	1599.5000	15.248821	0
				BLL13	355.8333	3.392334	0
				BLL14	730.8333	6.967394	0

Table 1.
Results of GLMM studies on roost-tier (branch length L1=BLL1 to BLL5 in Albizia lebbeck; *branch length L1=BLL1 to BLL14 in* Tamarindus indica *and in both sites BLL1 acts as intercept) preference in roost-trees of the bat* P. giganteus *in* A. lebbeck *at Joteghanashyam and* T. indica *at Simla, West Bengal, India.*

5. Discussion

Various workers [5, 6, 8, 9, 12–14] have paid due attention on the choice and selection of trees for roosting by the bats. Though they have paid due attention on DBH, canopy nature, foliages and age of the roost trees no information in respect to preference of roosting branches is on record. From the results it is evident that the bats *P. giganteus* have preference for roost sites in a roost tree. And, from the present findings it is clear that these bats have a priority to avail the opportunity to hang in the branches occurring at the mid sector of the tree. As they are colonial in habit and all the members of a colony are habituated to use the same roost tree if and when possible, the late comers have no alternative but to hang in the branches where spaces are available, even these branches being less preferred.

The bats *P. giganteus* left the roost tree at the onset of darkness to fly to the foraging sites. Depending upon the availability of food sources some individuals being well fed at the early hours may try to return the roost tree as early as possible, perhaps to take the shelter in the preferred branches of the roost tree. This kind of behavior most probably related with the assurance of individual's safety from the effect of adverse conditions viz. the attack by the predators [19], speed of the severe cyclonic wind [20] extremely high temperature and heat being exposed directly to the sunlight, direct hit of the rain drops, extreme cold waves during cooler months and loo during summer [21].

Thus, we hypothesized that the bats *P. giganteus* have preferred roost tiers in a roost-tree to ensure self protection through the exercise of their subtle intra-specific competition.

Acknowledgements

The authors are indebted to the Head of the Department of Zoology, Sidho-Kanho-Birsha University, Purulia and to the Head of the Zoology Department, University of Calcutta for the facilities provided. Also, the authors are thankful to the local people of the study areas for their cooperation.

Research funding

None declared.

Conflict of interest

There exists no conflict of interest.

Research ethics and best practices

The guidelines on Animal Ethics, supplied by the University Grants Commission, New Delhi were followed to carry out this work.

Author details

Susanta Mallick[1], Asif Hossain[1] and Srimanta Kumar Raut[2*]

1 Department of Zoology, Sidho-Kanho-Birsha University, Purulia, West Bengal, India

2 Ecology and Ethology Laboratory, Department of Zoology, University of Calcutta, Kolkata, West Bengal, India

*Address all correspondence to: srimantakraut@gmail.com

References

[1] Fenton, M. B. 1997. Science and the conservation of bats, Journal of Mammalogy, 78(1):1-14.

[2] Ali, A. 2010. Population trend and conservation status of Indian flying fox *Pteropus giganteus* Brunnich, 1782 (Chiroptera: Pteropodidae) in Western Assam, Ecoscan, 4: 311-312.

[3] Bhandarkar, Sudhir V. and Paliwal, Gopal T. 2018. Ecological notes on roosts of *Pteropus giganteus* (Brunnich, 1782) in eastern Vidarbha Maharashtra India. ESSENCE, International Journal for Environmental Rehabilitation and Conservation, IX (1): 67-70.

[4] Elangovan, V. and Kumar, M. 2019. Effect of tree characteristics on roost selection of the Indian flying fox, *Pteropus giganteus*, Journal of Bat Research and Conservation, 12(1): 100-106.

[5] Evelyn, M. J. and Stiles, D. A. 2003. Roosting requirements of two frugivorous bats (*Sturnira lilium* and *Artibeus intermedius*) in fragmented neotropical Forest, Biotropica, 35(3): 405-418.

[6] Gulraiz, T. L., Javid, A., Mahmood-Ul-Hassan, M., Maqbool, A., Ashraf, S., Hussain, M., and Daud, S. 2015. Roost characteristics and habitat preferences of Indian flying-fox (*Pteropus giganteus*) in urban areas of Lahore, Pakistan, Turkish Journal of Zoology, 39: 388-394. DOI:10.3906/zoo-1401-71.

[7] Hutchinson, J. and Lacki, M. 2000. Selection of day roosts by red bats in mixed Mesophytic forests, The Journal of Wildlife Management, 64: 87. DOI:10.2307/3802977.

[8] Jeyapraba, L. 2016. Roosting Ecology of *Pteropus giganteus* (Brunnich, 1782) Indian Flying Fox and Threats for their Survival, DOI:10.5281/zenodo.179134.

[9] Kalcounis-Rüppell, M. C., Psyllakis, J. M. and Brigham, R. M., 2005. Tree roost selection by bats: An empirical synthesis using meta-analysis, Wildlife Society Bulletin, 33: 1123-1132.

[10] Mager, Kenneth J. and Nelson, Thomas A. 2001. Roost site selection of the red bat (Lasiurus borealis), American Midland Naturalist, 145: 120-126.

[11] Nunes, H., Rocha, F. L. and Cordeiro-Estrela, P. 2017. Bats in urban areas of Brazil: Roosts, food resources and parasites in disturbed environments, Urban Ecosystems, 20:953-969.

[12] Dietz, M., Brombacher, M., Erasmy, M., Fenchuk, V. and Simon, O. 2018. Bat community and roost site selection of tree-dwelling bats in a well-preserved European lowland Forest, Acta Chiropterologica, 20. 117-127. DOI:10.3161/15081109ACC2018.20.1.008.

[13] Neupane, K. R., Basnet, K. and Katuwal, H. B. 2016. Plight of Indian flying fox (*Pteropus giganteus*) in lowlands of eastern Nepal, World Journal of Zoology, 11(2): 81-85. DOI:10.5829/idosi.wjz.2016.11.2.10355

[14] Silvis, A., W.M. Ford and E.R. Britzke. 2015. Day-roost tree selection by northern long-eared bats—What do random tree comparisons and one year of data really tell us? Global Ecology and Conservation, 3:756-763

[15] Brünnich, M. T. 1782. Dyrenes Historie or Dyre-Samlingen Udi Universitetets Natu-Theater. Vol. 1. Nicolaus Moller, Kiobenhavn.

[16] Tsang, S.M. 2020. "*Pteropus giganteus*". IUCN Red List of Threatened Species. 2020: E.T18725A22081011. DOI:10.2305/IUCN.UK.2020-3.RLTS. T18725A22081011.en. Retrieved 15 December 2020.

[17] Khan, Mohammad. 2001. Status and distribution of bats in Bangladesh with notes on their ecology, Zoos' Print Journal, 16(5): 479-483.

[18] Srinivasulu, C., Racey, P.A. and Mistry, S. 2010. A key to the bats (Mammalia: Chiroptera) of South Asia, Journal of Threatened Taxa, 2(7): 1001-1076. DOI:10.11609/JoTT.o2352.1001-76.

[19] Mallick, S., Hossain, A. and Raut, S.K. 2021. Predators of The Bat *Pteropus giganteus* Occurring In West Bengal, India, International Journal of Recent Scientific Research, 12(06):42138-42140. DOI:10.24327/ijrsr.2021.1206.6035.

[20] Mallick, S. and Raut, S. K. 2010. Effect of Aila storm on flying fox *Pteropus giganteus giganteus* (Brunnich), Journal of the Bombay Natural History Society, 107(2): 167.

[21] Mallick, S., Hossain, A. and Raut, S.K. 2020. On certain devices of the flying fox, *Pteropus giganteus* to overcome summer heat at Purulia, India, Proceedings of the Zoological Society of India, 19(1): 1-7. http://anveshika.org/proceedings-of-the-zoological-society-of-india/wp-content/uploads/sites/2/2020/06/1-JUNE-2020.